MEM05005B

2015

Carry out mechanical cutting

First Published March 2014

Edition 1 – March 2014

Edition 2 – October 2015

Conditions of Use:

Unit Resource Manual

Manufacturing Skills Australia Courses

This Student's Manual has been developed by BlackLine Design for use in the Manufacturing Skills Australia Courses.

Additional resource units can viewed and be ordered at www.acru.com.au

Feedback:

Your feedback is essential for improving the quality of these manuals.

This unit has not been technically edited. Please advise BlackLine Design of any changes, additions, deletions or anything else you believe would improve the quality of this Student Workbook. Don't assume that someone else will do it. Your comments can be made by photocopying the relevant pages and including your comments or suggestions.

Forward your comments to:

> BlackLine Design
>
> blakline@bigpond.net.au
>
> Sydney, NSW   2000

Corporate Licenses

State and National TAFE Colleges and Institutes, and Registered Training Organisations are eligible to purchase corporate licenses.

All licenses are perpetual and allow the licensee to upload the material onto a delivery system (Moodle etc), print the resource in book form and sell or distribute the material to enrolled students within their organisation. The license allows the holder to re-badge the material but must retain acknowledgment to BlackLine Design as the original developer and owner.

Aims of the Competency Unit:

This unit covers setting up and operating a range of mechanical cutting and holing equipment and applies to sawing, shearing, cropping and/or holing and includes setting up and operating a range of equipment. Examples of machines that could be covered include guillotines, croppers, cold saws, band saws, automatic saws etc. Typical applications of this unit may include cutting for manufacture, production cutting and cutting of materials selected from stores in a maintenance environment.

The unit does not cover hand or hand held power tools used for cutting e.g. circular saws, nibblers and side grinder. These skills are covered by Unit MEM18001C (Use hand tools) and Unit MEM18002B (Use power tools/hand held operations).

This unit does not include the skills required for operational maintenance of the equipment used; these skills are covered by Unit MEM07001B (Perform operational maintenance of machines/equipment).

For repair and welding of band saw blades where blade repair unit is not attached to the machine, refer to Unit MEM05013C (Perform manual production welding). .

Unit Hours:
18 Hours

Prerequisites:
MEM12023A - Perform engineering measurements.

MEM18001C – Use hand tools

Elements and Performance Criteria

1.	Determine job requirements	1.1	Job requirements and specifications are determined from job sheets and/or instructions.
		1.2	Appropriate method/machine is selected to meet specifications.
		1.3	Machine is loaded and adjusted for operation consistent with standard operating procedures.
2.	Select/set up machine tooling.	2.1	Tooling is selected to match job requirements.
		2.2	Report environmental incidents to appropriate personnel.
		2.3	Machine is set up and adjusted using standard operating procedures.
3.	Operate mechanical cutting machine	3.1	Appropriate stops and guards are set and adjusted as required.
		3.2	Make suggestions for improvements to workplace practices in own work area.
		3.3	Material is secured and correctly positioned using measuring equipment as necessary.
		3.4	Machine is operated to cut/hole material to specifications using standard operating procedures.
4.	Check material for conformance to specification	4.1	Material is checked against specification. Machine and/or tooling is adjusted as required and in process adjustments carried out as necessary.
		4.2	Material is cut and/or holed to within workplace tolerances.
		4.3	Material is used in most economical way.
		4.4	Codes and standards are observed.

Required Skills and Knowledge:

Required skills include:

- loading and adjusting cutting machines
- selecting machines and tooling
- installing cutting tool
- setting up and adjusting cutting machine
- securing and correctly positioning materials
- cutting and holing materials
- applying relevant codes and standards
- reading and interpreting routine information on written job instructions, specifications and standard operating procedures
- following oral instruction
- measuring materials to specified workplace tolerances and within the machine range
- clarifying routine task-related information

Required knowledge includes:

- the characteristics of cutting methods and machines
- effect of materials on the machine tooling, tooling defects and adjustments
- effect of adjustments on the dimensions of the cut material
- applicable tolerances
- methods of marking out materials to ensure minimum wastage
- any applicable industry standards, national/Australian standards, NOHSC guides, State/Territory regulatory codes of practice/standard
- use and application of personal protective equipment for mechanical cutting
- safe work practices and procedures

Lesson Program:

Unit hour unit and is divided into the following program.

Topic	Skill Practice Exercise
Topic 1 – Cutting Methods:	MEM05005-RQ-0101
Topic 2 – Metal Turning:	MEM05005-RQ-0201
Topic 3 – Drill Press:	MEM05005-RQ-0301
Topic 4 – Metal Shears	MEM05005-RQ-0401 MEM05005-SP-0402 MEM05005-SP-0403
Topic 5 – Cold Saw:	MEM05005-RQ-0501 MEM05005-SP-0502
Topic 6 – Band Saw:	MEM05005-RQ-0601 MEM05005-SP-0602
Practice Competency Test	MEM05005–PT-01

Contents:

Topic 1 – Cutting Methods:

Required Skills:
On completion of the session, the participants will be able to:

- Identify the metal, thermal and hydraulic methods of cutting metal.

Required Knowledge:
- Types of cutting processes.

1.1 Cutting Methods:
Mechanical cutting consists of a various number of processes in which a piece of raw material is cut into a desired final shape and size by a controlled material-removal process. Mechanical and laser cutting machinery are common fabricating processes used in today's manufacturing industries. Each method employs its own distinct equipment, and has its own advantages and disadvantages. Preference among the two usually depends on a range of factors, such as application requirements, cost-effectiveness, and production capabilities.

1.1.1 Mechanical Cutting:
Mechanical cutting, which includes tooling and machining, is a process that uses power-driven equipment to shape and form material into a predetermined design. Some common machines used in mechanical cutting include lathes, milling machines, drill presses, metal shears, croppers and saws and will be covered individually in the following topics.

1.1.2 Thermal Cutting:
Laser cutting uses an energy emission device to focus a highly-concentrated stream of photons onto a small area of a workpiece and cut precise designs out of the material. Lasers are typically computer-controlled and can make highly accurate cuts with a quality finish. The most common laser cutters are of the gaseous CO_2 or Nd:YAG variety.

1.1.2.1 Gas:
Oxyacetylene is a process that uses fuel gases and oxygen to cut metal. Pure oxygen, instead of air, is used to increase the flame temperature to allow localized melting of the workpiece material (e.g. steel) in a room environment. A common propane/air flame burns at about 2,000 °C, a propane/oxygen flame burns at about 2,500 °C, and an acetylene/oxygen flame burns at about 3,500 °C.

Oxy-fuel is one of the oldest welding processes, besides forge welding and is still used in industry; in recent decades it has been less widely utilized in industrial applications as other specifically devised technologies have been adopted. It is still widely used for welding pipes and tubes, as well as repair work and is also suitable, and favoured, for fabricating particular types of metal-based artwork. Oxy-fuel has an advantage over electric welding and cutting processes in situations where accessing electricity (e.g., via an extension cord or portable generator) would present difficulties; it is more self-contained, or, "more portable".

In oxy-fuel cutting, a torch is used to heat metal to its kindling temperature. A stream of oxygen is then trained on the metal, burning it into a metal oxide that flows out of the kerf as slag.

1.1.2.2 Electrical Gas Discharge:
Electrical Gas Discharge is a manufacturing process where a shape is obtained using electrical discharges or sparks. Material is removed from the workpiece by a series of rapidly recurring current discharges between two electrodes, separated by a dielectric

liquid and subject to an electric voltage. One of the electrodes is called the tool-electrode, or simply the 'tool' or 'electrode', while the other is called the workpiece-electrode, or 'workpiece'.

Decreasing the distance between the two electrodes reduces the intensity of the electric field in the volume between the electrodes becomes greater than the strength of the dielectric which breaks, allowing current to flow between the two electrodes. This phenomenon is the same as the breakdown of a capacitor; as a result, material is removed from both the electrodes. Once the current flow stops (or it is stopped – depending on the type of generator), new liquid dielectric is usually conveyed into the inter-electrode volume enabling the solid particles (debris) to be carried away and the insulating properties of the dielectric to be restored. Adding new liquid dielectric in the inter-electrode volume is commonly referred to as flushing; also, after a current flow, a difference of potential between the two electrodes is restored to what it was before the breakdown, so that a new liquid dielectric breakdown can occur.

1.1.2.3 Beams:
Laser beam machining involves using laser beam technology to perform functions typically accomplished by conventional cutting machines. The type of lasers most often used include the carbon dioxide (CO_2) and the neodymium doped:yttrium aluminium garnet (Nd:YAG). The adaptability of these tools allows them to perform more than one function, and the wide range of industries that often use laser beam machining technology includes automakers and jewellers.

A CO_2 laser is one of the more powerful types of laser used in laser beam machining. These lasers can generate 400 to 1,500 watts of power, which can cut through 25 mm thick carbon steel. The tool uses mirrors that direct the proton laser beam to the desired cutting location. The laser generally makes a tapered cut as it moves along the z-axis while the work surface travels along the x and y-axes. Industries generally use the power of the CO_2 laser for cutting and profiling.

The flexibility of the YAG laser beam enables manufacturers to use a machine that transmits the beam directly to the cutting surface or through something as small as a fibre optic cable. Lasers transmitted through fibre optics can be incorporated into robotic machines that can move on any axis around a stationary work site. While not as powerful as a CO_2 laser, a YAG laser can drill a hole to a depth of six times the diameter of its beam. Besides laser boring, industries commonly employ YAG lasers for etching and engraving.

Depending on the function required, industries utilize CO_2 or YAG tools for laser beam machining, and computer numerical control (CNC) instrumentation relays desired tasks to the laser. Manufacturers design each tool in sizes ranging from tabletop models to free standing room-sized machines. Small business owners and large industrial factories both use laser beam machining on materials ranging from cardboard, cork, and wood to steel, steel alloys, and stone.

Industrial manufacturing applications include cutting or welding metals in aircraft, automotive, and shipbuilding factories. Jewellers also use laser welding on delicate pieces of jewellery, and machinists use laser beam machining to resurface corroded parts by fusing material to damaged areas. Laser beams perform intricate cuts in plastic and metal sheeting for components installed in household electronics or machinery. Functioning similarly to an ink jet printer, laser beam machining is also used to engrave glass, plastic, and stone.

1.1.3 Hydraulic Cutting:
Hydraulic cutting is also known as water jet and is an industrial tool capable of cutting a wide variety of materials using a very high-pressure jet of water, or a mixture of water and an abrasive substance. The term abrasive jet refers specifically to the use of a mixture of water and abrasive to cut hard materials such as metal or granite. The terms pure water jet and water-only cutting refer to water jet cutting without the use of added abrasives and is often used for softer materials such as wood or rubber.

Topic 1 – Cutting Methods

Water jet cutting is often used during fabrication of machine parts. It is the preferred method when the materials being cut are sensitive to the high temperatures generated by other methods. Water jet cutting is used in various industries including mining and aerospace for cutting, shaping, and reaming.

The water jet cutter is commonly connected to a high-pressure water pump where the water is then ejected from the nozzle, cutting through the material by spraying it with the jet of high-speed water. Additives in the form of suspended grit or other abrasives, such as garnet and aluminium oxide, can assist in this process.

An important benefit of the water jet is the ability to cut material without interfering with its inherent structure, as there is no "heat-affected zone". Minimizing the effects of heat allows metals to be cut without harming or changing intrinsic properties. Water jet cutters are also capable of producing intricate cuts in material; with specialized software and 3-D machining heads, complex shapes can be produced.

The kerf, or width, of the cut can be adjusted by swapping parts in the nozzle, as well as changing the type and size of abrasive. Typical abrasive cuts have a kerf in the range of 1.016 to 1.27 mm, but can be as narrow as 0.508 mm. Non-abrasive cuts are normally 0.178 to 0.33 mm, but can be as small as 0.076 mm, which is approximately that of a human hair; these small jets can permit small details in a wide range of applications. Water jets are capable of attaining accuracies down to 0.13 mm and repeatability down to 0.025 mm.

Due to its relatively narrow kerf, water jet cutting can reduce the amount of scrap material produced, by allowing uncut parts to be nested more closely together than traditional cutting methods. Water jets use approximately one half to one gallon per minute (depending on the cutting head's orifice size), and the water can be recycled using a closed-loop system. Waste water usually is clean enough to filter and dispose of down a drain. The garnet abrasive is a non-toxic material that can be recycled for repeated use; otherwise, it can usually be disposed in a landfill. Water jets also produce fewer airborne dust particles, smoke, fumes, and contaminants, reducing operator exposure to hazardous materials.

Skill Practice Exercises:

Skill Practice Exercise MEM05005-RQ-0101

Investigate the cutting equipment and itemise in three columns the different types of metal cutting machinery in your workplace:

Mechanical	Thermal	Hydraulic

Topic 2 – Metal Turning:

Required Skills:
- Identify the lathes, mills and grinders used in cutting metal.
- Select a suitable machine to suit a specific task.
- Select the correct Personal Protective Equipment to use with individual pieces of machinery.

Required Knowledge:
- Reading and interpreting instructions.
- Application of various metal turning machines.
- Workplace Health and Safety Regulations.

2.1 Introduction:
This topic is intended as an introduction and appreciation to the types of metal turning machinery. Precise instructions on the setting up and operation of the individual types of machines are covered in specific MEM units.

Several types of metal turning machinery are available for use in industry; most are used mainly in dedicated Fitting and Machining Workshops but are not only restricted to those premises. Operators of metal turning machines require special training to operate the machines safely and efficiently. Specific competency units are provided for the training in the use of these machines; this topic is intended as an introduction to the most commonly used timber and metal working machines and as such only covers the types of machines and their applications.

2.2 Lathes:
There are several different types of lathe machinery designed for various purposes. A timber lathe is a small, simple device used exclusively for working with timber materials. An engine lathe is designed for common metalworking tasks while a turret lathe can perform multiple functions on the same work piece. Tool room lathe machinery is typically utilized for procedures requiring a high degree of accuracy. Computer numerically controlled lathe machinery is capable of performing high-speed functions for mass-production applications.

Lathe machinery is utilized to perform a process known as turning; this process involves removing a certain amount of material from a workpiece by means of rotation. A cutting tool is pressed against the workpiece to shape it as it turns on the lathe. There are several different types of lathe machinery intended for specific purposes. Most lathes are manually operated, but automated versions are also manufactured.

2.2.1 Wood Lathe:
A wood lathe is a machine used to smooth and shape wood and consists of a tool rest, headstock, tailstock, and bed and is not designed for precision cutting because the shaping tool is hand-powered. Spinning is controlled by a belt-driven motor which can be adjusted to various speeds. Woodworking tools are supported by the tool rest, and manual pressure is applied to the workpiece. A certain amount of skill is required to accurately operate a wood lathe.

In order to use a wood lathe, the crafter must insert a piece of timber into a special holder called a centre; this centre holds the wood in place so it can be shaped using the device with a variety of handheld blades. Some types of lathes, however, are capable of holding the blade in place, so the user does not have to hold the blade. The centre spins the wood rapidly as the crafter applies the blade to the surface of the workpiece.

Figure 2.1

A wood lathe can be fitted with a variety of different blades which can be of different widths, angles, and shapes. The type of blade determines the way the machine will cut the wood. Some blades are wide with sharper angles and intended to cut large portions of wood away from the piece. Other blades are sharper or pointier, and are used for detailing.

An operator may also use a wood lathe to assist with sandpapering the surface of the wood. Rather than holding the piece in hand and rubbing it repeatedly with a piece of sandpaper, the operator can allow the machine to spin the piece as they hold a piece of sandpaper against it; this makes it possible to finish sanding in a much shorter period of time and is also less physically demanding on the user.

Wood lathes can be used to create a variety of objects. For crafting and general woodworking, they can create decorative spindles to be used in the creation of table and chair legs; they can also be used to make cue sticks, baseball bats, and any other wooden object that needs to be cylindrical and smooth in shape.

2.2.1.1 Personal Protection Equipment:

 Approved safety glasses must be worn in all workshops.

 Long and loose hair must be contained at all times in all workshops.

 Approved footwear with substantial uppers must be worn in all workshops.

 Close fitting clothing or overalls must be worn in all workshops.

 Hearing protection should be worn when noise levels are excessive.

 A mask must be worn when excessive airborne dust is created.

 Exposed rings and jewellery must be removed.

2.2.2 Engine Lathe:

An engine lathe is designed for turning almost any type of metal work piece. Like the wood lathe, it consists of a headstock, tailstock, and bed; shaping tools may be manually operated or gear driven. An engine lathe also features a compound rest and cross-slide for angular and cross cutting; this machine typically has a much wider speed range than a wood lathe.

The engine lathe is a horizontally shaped piece of machinery that is most often used to turn metal manually. By turning the metal and using special cutting tools, the lathe is capable of forming the metal into specific shapes. As its name implies, it is often used to create metal pieces for use in an engine, whether it be for an automobile, a tractor, a boat, or any other motorized vehicle or machine.

Although people use the engine lathe primarily for spinning sheet metals, it is also used for drilling, making square blocks, and creating shafts. Candle cup dies, forge burner nozzles, foundry pattern core boxes, and prints are other products it can produce. Most modern tools were made with the help of an engine lathe.

Figure 2.2

The features of an engine lathe include gears, a carriage, a tailstock, and a stepped pulley used for various spindle speeds. The gears are used to power the carriage. In turn, the carriage bolsters the cutting tools. The tailstock is used to support the hole-drilling process that takes place in the spindle.

2.2.2.1 Personal Protection Equipment:

 Approved safety glasses must be worn in all workshops.

 Long and loose hair must be contained at all times in all workshops.

 Approved footwear with substantial uppers must be worn in all workshops.

 Close fitting clothing or overalls must be worn in all workshops.

 Gloves must not be worn when using this machine.

 Exposed rings and jewellery must be removed.

2.2.3 Turret Lathe:

A turret lathe is used by metalworking shops to fashion parts that are a standard size and shape; this style of lathe has an indexed tool holder, which means that a series of cuts may be made to the piece in succession. The operator does not need to stop in between cuts to set up for the next one. The path used to make the metal parts using a turret lathe is controlled by the machine itself, which leads to a more efficient process.

The turret lathe has been in use since the mid-19th century and the development was an important one for manufacturers. Before the turret lathe came into existence, making quality metal tools or components was dependent on the skill of the operator. Once it started being used in manufacturing plants, it meant that tools and other parts could be made quicker and at a lower cost.

When mass production is the required, this type of lathe is the most efficient choice for producers. The settings for each type of tool can be stored. Changing the settings when it's time to produce a different part is a quick and easy process.

Figure 2.3

2.2.3.1 Personal Protection Equipment:

 Approved safety glasses must be worn in all workshops.

 Long and loose hair must be contained at all times in all workshops.

 Approved footwear with substantial uppers must be worn in all workshops.

 Close fitting clothing or overalls must be worn in all workshops.

 Hearing protection should be worn when noise levels are excessive.

 Gloves must not be worn when using this machine.

 Exposed rings and jewellery must be removed.

2.2.4 Spinning Lathe:

A spinning lathe is a lathe that forms thin-walled ductile metals into a variety of cylindrical or semi-cylindrical shaped parts; this type of lathe has been derived from both the wood and engine lathes and, if tooled properly, each style of machine is capable of spinning. A spinning lathe has all of the major components of a wood lathe or engine lathe, but the types of tooling that are used and how these are mounted are different. Wood lathes or engine lathes typically are designed to cut material, but a spinning lathe is designed to form material.

Exterior tools, such as harden rollers or forming spoons, are used to force the spinning metal into the desired size and shape. Sometimes, a mandrel might also be employed to help form the interior of the spun metal shape. The spinning lathe makes a variety of products, from metal cones to compressed gas cylinders.

For a spinning lathe to work properly, the material to be spun must be properly chosen. A variety of metals can be chosen, but each of these should have a high ductility. Some metals that have been spun are aluminium, brass, copper, stainless steel and steel.

Figure 2.4

After a metal has been chosen, it is then important for the lathe operator to determine the initial shape and the final shape of the material. The initial and final shape will help the operator determine the size of the spinning lathe to be used. For instance, if a metal disc is spun, then a lathe with a large swing should be used. If the material is smaller or possibly cylindrical in shape, the swing distance might be reduced.

A spinning lathe is used for production and one-off parts. Spinning lathes set for production will generally have standardized tooling and might have a fast changeover capability. Individual parts will not necessarily be used with a dedicated spinning lathe, and the tooling might be of a more generic nature.

With spinning lathes, operations might be performed with a computer numerical control (CNC) or manually. Production spinning of very large parts might generally be paired with a CNC machine. When small quantities or relatively small parts are needed, a skilled lathe operator might produce such items.

For some spinning lathe operations, it is beneficial to heat the material as it is being formed. Heat naturally builds while a product is being spun. If the material is thicker or susceptible to stress fracturing, additional heat might be applied to assist the flow of material as the lathe is in operation.

2.2.4.1 Personal Protection Equipment:

 Approved safety glasses must be worn in all workshops.

 Long and loose hair must be contained at all times in all workshops.

 Approved footwear with substantial uppers must be worn in all workshops.

 Close fitting clothing or overalls must be worn in all workshops.

 Hearing protection should be worn when noise levels are excessive.

 Gloves must not be worn when using this machine.

 Exposed rings and jewellery must be removed.

2.2.5 CNC Machine:

Computer Numerical Control (CNC) machines are automated milling devices that make industrial components without direct human assistance by using coded instructions that are sent to an internal computer, which allows factories to fabricate parts accurately and quickly. There are many different types of CNC machines, ranging from drills to plasma cutters, so they can be used to make a wide variety of parts. Though most are used industrially in manufacturing, there are also hobby versions of most of the machines that can be used in private homes.

Figure 2.5

2.2.5.1 Types of Machines

The most common CNC machines are milling machines, lathes, and grinders. Milling machines automatically cut materials, including metal, using a cutting spindle, which can move to different positions and depths as directed by the computer instructions. Lathes use automated tools that spin to shape material; they're commonly used to make very detailed cuts in symmetrical pieces, like cones and cylinders.

Grinders use a spinning wheel to grind down materials, and mould metal or plastic into the desired shape. Grinders are easy to program, so they're usually used for projects that do not require the same precision as mills or lathes. Besides these, there are also CNC routers, which are used to make cuts in a variety of materials; as well as computer programmable 3D printers; and turret punches, which are used to make holes in metal or plastic. The technology can also be used with different types of cutters, including those that work with water, lasers, and plasma.

2.2.5.2 Programming and Operation

The code used to program CNC units is generically called G-Code. It contains information about where parts of the machine should be positioned, and tells the machine exactly where to place a tool. Other instructions tell the machine additional details, like the speed a part should run at; how deep it should cut, burn, or punch; and the angle of an automated tool. Most modern industrial CNC machines are tied into a network of computers, and receive operating and tooling instructions via a software file.

2.2.5.3 Advantages and Disadvantages

In an industrial setting, CNC machines can be combined into entire cells of tooling machines that can operate independently of each other; they are often driven by completely digital designs, which eliminate the need for design blueprints to be physically drawn up. Many are capable of running for several days without human intervention. In fact, some are so sophisticated that they can contact the operator's cell phone and send an alert if a malfunction occurs. The automated features make it possible to produce thousands of parts with minimal supervision, and free the operator to perform other tasks.

Besides this, a CNC machine can form parts with a level of precision that is nearly impossible using older tools. In a conventional factory, workers must control different tools by hand, and errors are common, but a machine can perform the same task without becoming tired, and can work non-stop which saves a lot of time, and the improved accuracy can help eliminate waste, since there are less faulty parts that have to be thrown away.

Despite their advantages, CNC machines are more expensive than older types of machines, which can make them unaffordable for smaller operations. They're also expensive to repair and maintain. Also, though they do limit the potential for errors, they don't eliminate it entirely, since operations can still program or operate the machine incorrectly. Additionally, these machines need to be operated by a skilled workforce with a specific type of training, which may not be available in all areas.

2.2.5.4 Development

CNC machines have evolved considerably since their initial introduction into the manufacturing industry. The earliest CNC machines received code instructions through hard-wired controllers, which meant that the programming format could not be altered. Later models were programmed via mainframe cables and floppy disks, which permitted variations in programming. Modern CNC machines can be operated by software files found on CDs, USB drives, or sent over a network.

2.2.5.5 Personal Protection Equipment:

 Approved safety glasses must be worn in all workshops.

 Long and loose hair must be contained at all times in all workshops.

 Approved footwear with substantial uppers must be worn in all workshops.

 Close fitting clothing or overalls must be worn in all workshops.

 Exposed rings and jewellery must be removed.

2.3 Milling Machine:

A milling machine is a tool found primarily in the metalworking industry. In general, these machines are used to shape solid products by eliminating excess material in order to form a finished product. Milling machines can be used for a variety of complicated cutting operations – from slot cutting, threading, and rabbeting to routing, planning, and drilling. They are also used in die sinking, which involves shaping a steel block so that it can be used for various functions, such as moulding plastics or coining.

A milling machine is usually capable of cutting a wide variety of metals, ranging from aluminium to stainless steel. Depending on the material being cut, the machine can be set to move at a faster or slower pace. Softer materials are generally milled at higher speeds while harder materials usually require slower speeds. In addition, harder materials often require smaller amounts of material to be milled off at one time.

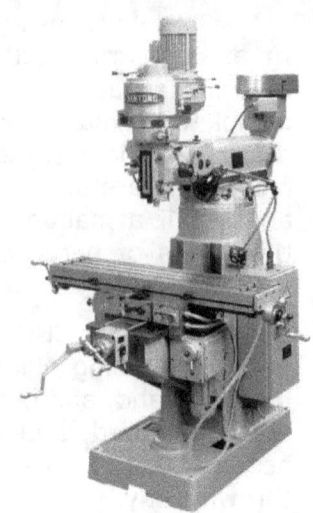

Generally, a milling machine operator runs the machine by feeding the material over a device called a milling cutter. As the material is fed past the cutter, the cutter's teeth slice through the material to form the desired shape. Using gadgets like precision ground slides and lead screws, the movement of the material as well as the cutter can be kept to less than 0.001 inches (about 0.025 millimetres) in order to make the cut exact.

Figure 2.6

In addition to a cutter, a typical mill machine contains a spindle axis, which is a device that holds the cutter in place. The cutter revolves around the spindle axis, and the axis can usually be adjust to varying speeds. Most machines also come with a worktable that can be used to support and feed the material. The worktable generally moves in two directions, and most modern worktables are power-operated. Additionally, a modern milling machine is typically equipped with a self-contained electric drive motor and a coolant system.

From micro, mini, and benchtop to floor standing, large, and gigantic, a milling machine can be found in a variety of sizes. Milling machines can have flat, angular, curved, or irregular surfaces. In addition, they can have a vertical or a horizontal orientation. A vertical milling machine has a spindle axis that faces vertically while a horizontal machine's spindle faces horizontally.

Milling machinery can be operated manually or digitally using device called a computer numerical control or CNC milling machine. In addition to the traditional X, Y, and Z axes found in a manual machine, a milling CNC machine often contains one or two additional axes. These extra axes can allow for greater flexibility and more precision. CNC machines eliminate the need for a machine operator, which can prevent possible accidents as well as save on labour costs.

2.3.1 Personal Protection Equipment:

 Approved safety glasses must be worn in all workshops.

 Long and loose hair must be contained at all times in all workshops.

 Approved footwear with substantial uppers must be worn in all workshops.

 Close fitting clothing or overalls must be worn in all workshops.

 Hearing protection should be worn when noise levels are excessive.

 Gloves must not be worn when using this machine.

 Exposed rings and jewellery must be removed.

2.4 Grinder:

A universal grinder is a tool that grinds, files and finishes workpieces, most of which are made of metal. Universal grinders are different from most other types of grinders because their construction and arrangement allows them to work with an incredibly wide variety of materials and tools that require grinding processes. Most grinders manufactured are material-specific, meaning that they can only work on certain types of materials. For example, a metal grinder is different from a tool grinder, mainly because of the variance in grit, size and shape. All grinders, universal or not, work in a similar manner.

Grinders work by spinning a coarse surface at very high speeds, which slowly cuts away at small pieces on the applied material. They are typically used for finishing applications, where the surface or edge of a workpiece after undergoing its initial process needs finishing. One example of this would be a split piece of wood. Oftentimes, the new surface resulting from splitting the wood is very rough, and requires either a planer or a belt sander, another type of grinder, to give it a smooth finish.

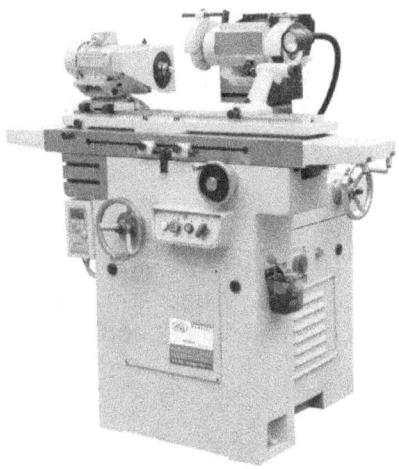

Most grinders are, however, used for metalworking applications. Metalworking processes outside of careful machining usually produce rough, uneven surfaces that require some sort of finishing technique.

Figure 2.7

An example of this would be plasma cutting metal, which while precise, often results in an uneven surface surrounding the cut end. While most flat surfaces or ends could undergo a milling process, such precise machining is often unnecessary for many jobs. As a result, a grinder or universal grinder is employed, which can give the workpiece a reasonably smooth finish that is adequate for most purposes.

The universal grinder can handle all of the aforementioned jobs and many more. These grinders can be very useful for facilities which work with a wider variety of materials. For those kinds of facilities, a universal grinder is a worthwhile investment, as it eliminates the need to purchase individual grinders for a certain material or process. This allows for workers to grind anything from metals to tools that require regular sharpening.

Since there are many different types of grinders, a universal grinder needs to incorporate as many varieties as possible. This can be done in many ways, as all of them require some sort of rotary movement, typically performed by an electric motor. As for changing grinder types to suit different materials, this can be done through changing the grit count or grinder shape, or even adjusting the torque and speed of rotation.

2.4.1 Personal Protection Equipment:

 Approved safety glasses must be worn in all workshops.

 Long and loose hair must be contained at all times in all workshops.

 Approved footwear with substantial uppers must be worn in all workshops.

 Close fitting clothing or overalls must be worn in all workshops.

 Hearing protection should be worn when noise levels are excessive.

 Exposed rings and jewellery must be removed.

Skill Practice Exercises:

Skill Practice Exercise MEM05005-RQ-0201

Answer the following questions:

1. Which type of lathe would be best suited to manufacture 2,000 of the Cylinder Controller Head shown in image RQ1?

   _____

2. Which machine would be needed to manufacture the tin Spittoon indicated in image RQ2?

   A – Wood lathe

   B – Engine Lathe

   C – Turret Lathe

   D – Spinning Lathe

3. What lathe would be best suited to be used where a series of cuts can be made in succession?

   _____

4. Which machine would be best suited to create a keyway in a shaft as shown in image RQ4?

   _____

5. List the 2 main disadvantages for a machine shop purchasing a CNC machine.

   _____

   _____

6. Select the PPE that is not worn when operating an engine lathe.

A.   B.   C.   D.

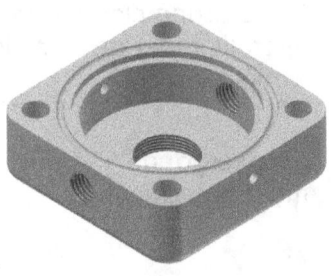   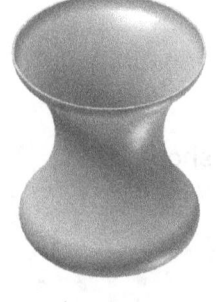   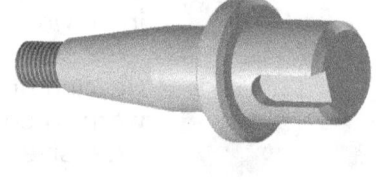

RQ1   RQ2

RQ4

Topic 3 – Drill Press:

Required Skills:

- Identify the different types of drills used in drilling holes.
- Select a suitable drill and bit to suit a specific task.
- Drill specified holes in various materials in accordance with instructions.
- Select the correct Personal Protective Equipment to use with individual pieces of machinery.

Required Knowledge:

- Reading and interpreting instructions.
- Use of various drills.
- Workplace Health and Safety Regulations.

3.1 Introduction:

A drill press (also called a Pedestal Drill) is a fixed style of drill that uses a motor-driven, rotating tool to drill holes in workpieces which generally consist of metal, timber and plastics. A drill press is an upright-standing fixed drill that can be mounted on a cabinet or to the floor and are available in different styles and types including Floor (or Standing), Bench Top, Upright Sensitive, Multi Spindle and Radial Arm.

A drill press is preferable to a hand drill when the location and orientation of the hole must be controlled accurately. A drill press is composed of a base that supports a column; the column in turn supports a table. Work can be supported on the table with a vise or hold down clamps, or the table can be swivelled out of the way to allow tall work to be supported directly on the base. Height of the table can be adjusted with a table lift crank than locked in place with a table lock. The column also supports a head containing a motor. The motor turns the spindle at a speed controlled by a variable speed control dial. The spindle holds a drill chuck to hold the cutting tools (drill bits, centre drills, deburring tools, etc.)

3.2 Drill Presses

Drilling machines, or drill presses, are primarily used to drill or enlarge a cylindrical hole in a workpiece or part. The chief operation performed on the drill press is drilling, but other possible operations include: reaming, countersinking, counterboring, and tapping.

The floor type drill press is found in both home and industrial workshops; this style drill press is composed of four major groups of assemblies: the head, table, column, and base. The head contains the motor and variable speed mechanism used to drive the spindle. The spindle is housed within the quill, which can be moved up or down by either manual or automatic feed. The table is mounted on the column, and is used to support the workpiece. The table may be raised or lowered on the column, depending upon the machining needs. The column is the backbone of the drill press. The head and base are clamped to it, and it serves as a guide for the table. The cast-iron base is the supporting member of the entire structure.

Topic 3 – Drill Press

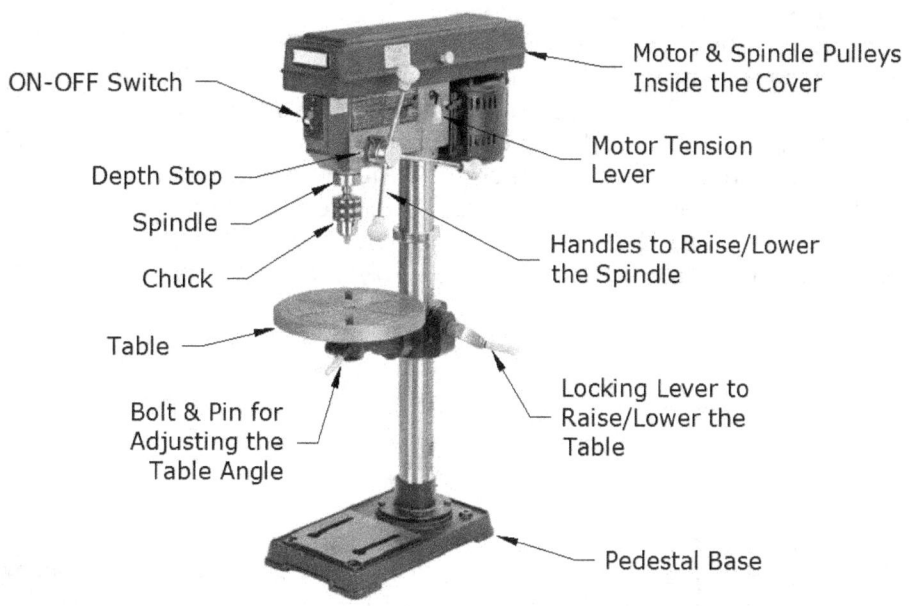

ON-OFF Switch

Depth Stop

Spindle

Chuck

Table

Bolt & Pin for
Adjusting the
Table Angle

Motor & Spindle Pulleys
Inside the Cover

Motor Tension
Lever

Handles to Raise/Lower
the Spindle

Locking Lever to
Raise/Lower the
Table

Pedestal Base

Figure 3.1

3.2.1 Pedestal Drill Press:

The Pedestal Drill Press is one of the most frequently used machine tools and are available as Floor (or Standing) and Bench mounted models; they are used mainly for drilling holes but reaming, countersinking, and boring can also be accomplished with the drill press.

Injuries on this equipment generally arise from careless operation by users who fail to take the time to clamp their work to the table or who try to hold the drill press vise in one hand while advancing the spindle in the other. Be sure to clamp all work securely before drilling. Another common injury results from broken cutting tools. When a drill begins to emerge from the bottom of the work piece on through-holes, the feed rate needs to be reduced in order to prevent the drill flutes from biting too aggressively into the work; this action sometimes causes cutting tool breakage. Reduce the feed force and let the drill advance slowly.

Most machine shops are equipped with at least one pedestal-type drill press; these machines are used to drill holes in wood, plastic, aluminium, brass, steel, and most other common engineering materials.

Figure 3.2

The speed control can have variable or multiple positions depending on the quality, make and model. Low range is used when larger holes or harder materials need to be machined while high range is useful when small holes and softer materials are being drilled. Some machines may have separate switches for power and range.

3.2.2 Upright Sensitive Drill Press:

The upright sensitive drill press (Figure 3.3) is a light-duty type of drilling machine that normally incorporates a belt drive spindle head. This machine is generally used for moderate-to-light duty work. The upright sensitive drill press gets its name due to the fact that the machine can only be hand fed. Hand feeding the tool into the workpiece allows the operator to "feel" the cutting action of the tool. The sensitive drill press is manufactured in a floor style or a bench style.

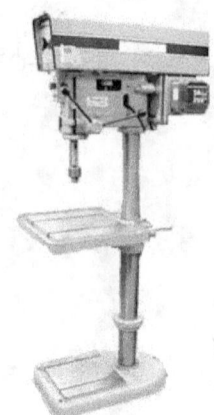

Figure 3.3

3.2.3 Multispindle Drill Press:

A multispindle drill press is a floor-type drill press used to handle a variety of jobs. Multspindle drill presses are used to drill workpieces with multiple holes at the same time. The drills are mainly used in mass production for machining workpieces that require simultaneous drilling, reaming and tapping of a large number of holes in different planes of the workpiece. A single drilling spindle machine is not economical for such purposes, as not only a considerably large number of machines and operators are required, but also the machining cycle is longer.

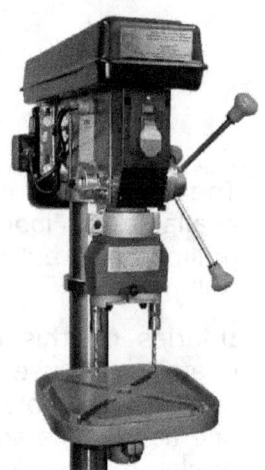

Figure 3.4

3.2.4 Radial Arm Drill Press:

A radial drill press is a machine tool that features an extended arm or beam along which a drill head can be moved to conveniently drill holes in large or cumbersome work pieces. A radial drill press allows the operator to advance the drill head along the machine's extended arm, rotate the arm in an arc, or tilt the drill head to drill at an angle. There are a wide range of radial drill press sizes available; big machine shop examples typically featuring large drill size capacities and drill bit cooling systems.

A drill press is a tool that allows for extreme precision in drilling operations. In a typical drill press operation, the drill head moves up and down to drill holes using a rotary handle and a geared bar. This allows the drill head to move vertically while maintaining a perfectly level, upright orientation. The result of this arrangement is added power on the drill stroke because of the mechanical advantage involved and the drilling of perfectly straight holes. The drawback with this system is the small work area which requires constant moving of the work piece if multiple holes require drilling; it also limits the possible size of the work piece.

A radial drill press resolves these challenges by allowing drill head movement in several additional directions while at the same time maintaining vertical alignment. This flexibility is achieved by including an elongated beam or carriage along which the drill head may be moved. This carriage can also rotate in an arc around the drill presses main upright member, thereby allowing for a very large area to be covered. In addition, most radial drill presses allow for the drill head to be tilted and then locked to drill holes in non-vertical orientations. This extended flexibility of motion allows for multiple holes drilled in very large work pieces without the need to constantly move the work pieces.

Heavy, industrial radial drill press machines are typically fitted with chucks capable of accepting very large drill bits with 100 mm capabilities being common on larger examples. Provision is also commonly made for drill bit cooling with the fitment of cooling fluid tanks, pumps, and nozzles. Smaller do-it-yourself radial drill presses are also available and feature most of the flexibility of their larger, machine shop siblings. These smaller units usually feature table top mounting and standard drill press vice tables.

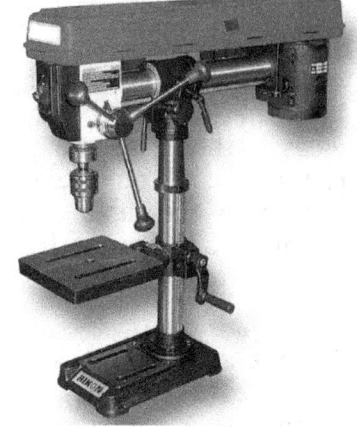

Figure 3.5

3.2.5 Gang Drill Press:
The gang style drilling machine (Figure 3.6) or gang drill press has several work heads positioned over a single table. This type of drill press is used when successive operations are to be done. For instance, the first head may be used to spot drill. The second head may be used to tap drill. The third head may be used, along with a tapping head, to tap the hole. The fourth head may be used to chamfer.

Figure 3.6

3.2.6 Micro Drill Press:
The micro drill press is an extremely accurate, high spindle speed drill press. The micro drill press is typically very small (Figure 3.7) and is only capable of handling very small parts. Many micro drill presses are manufactured as bench top models. They are equipped with chucks capable of holding very small drilling tools.

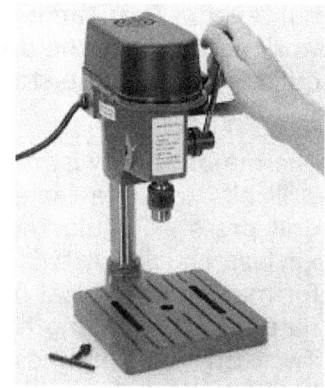

Figure 3.7

3.2.7 Turret Drilling Machines:
Turret drilling machines are equipped with several drilling heads mounted on a turret (Figure 3.8). Each turret head can be equipped with a different type of cutting tool. The turret allows the needed tool to be quickly indexed into position. Modern turret type drilling machines are computer-controlled so that the table can be quickly and accurately positioned.

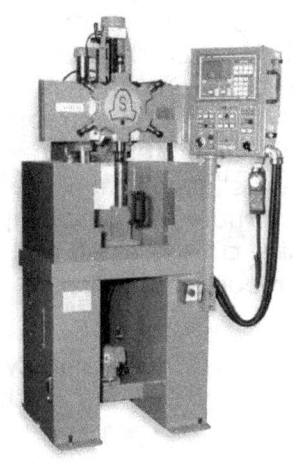

Figure 3.8

3.3 Drill Accessories:
3.3.1 Drill Vice:

A drill press vise is a device designed to secure a workpiece firmly enough between its jaws that it won't move at all when the drill bit is pressed into it. It's similar to other vices in that it uses a threaded rod to force two jaws together. Some drill press vices clamp to the table itself, while others have systems that use the mitre-gauge slot to secure to the table. One notable drawback is that most drill press vices have jaw openings of no more than 150 mm to 180 mm, requiring larger workpieces to be secured by other methods.

There are two ways to secure a workpiece firmly enough to prevent most movement. The first is to use clamps to secure the workpiece to the table, and the second is to use a drill press vice. Drill press clamps are preferable for larger or irregularly-shaped workpieces, but won't work well for thicker or tall pieces, such as when the core must be drilled out of a lamp base. Drill press vices are best suited for these pieces as well as for narrower workpieces that will fit within the relatively narrow space between the vice's jaws.

Figure 3.9

A drill press vice is very well-suited for some forms of production work where multiple holes must be drilled. Once the location of the first hole is determined, the vice is set in place and tightened, and the hole is drilled. The vice is loosened and the workpiece is moved so that the next hole location is lined up under the drill bit. At this point, all that's necessary is to ensure that the drill bit is lined up properly and the vice is again tightened. The same concept holds true when drilling holes in multiple identical workpieces; once the drill press vice has been secured to the work table, the work can go quicker because the stationary jaw of the vice acts much like a fence.

3.3.2 Twist Drill Bits:

The majority of twist drill bits are made of high speed steel. Drill bits that are tapered, fit directly in the spindle of the drill press. Regular non-tapered bits, sit into a keyed or keyless chuck. A twist drill is a pointed cutting tool used for making cylindrical holes in the workpiece. It has helical flutes along its length for clearing chips from the holes. Twist drills are the most common used today, but there are many other styles with different purposes.

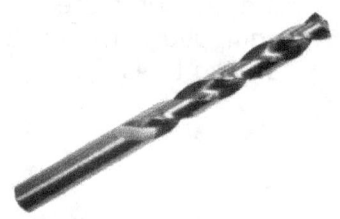

Figure 3.10

A twist drill is composed of three major parts: a shank, body, and point. The shank is the part of the drill bit held in the spindle of the drill press. The drill press power is transferred through the shank. Shanks are either one of two styles, straight or tapered. Straight shank drills are held in a friction chuck. Slippage between the drill bit and the chuck is often a problem, especially for larger drills. When using drill bits larger than 12 mm diameter, tapered shank drill bits are often used; these provide greater torque with less slippage than straight shank drill bits. The body, as described above, generally has two flutes to clear chips; these flutes are not cutting edges and should not be used for side cutting as an end mill. The point of the drill bit does all of the cutting action, which produces the cut chips. The point is ground on the end of the drill bit.

Holes produced by twist drill bits are generally oversize by as much as up to 1% of the bit's dia. The accuracy of the hole is dependent on the following factors:
- size of the bit,
- accuracy of the bit's point,
- accuracy of the chuck,
- accuracy and rigidity of the spindle,

- rigidity of the press,
- rigidity of the workpiece in its setup.

All holes to be drilled should be started with a centre-punch, centre-drill, or both.

3.2.2.1 Auger Drill Bit:

The auger drill bit is intended to be used in timber and is equipped with a very thick and deep spiral flute along the twist of the bit. The deep flute is designed to remove the chips from the hole being drilled in a much more effective manner than a typical twist drill bit. Originally designed for use only in a hand brace, the modern auger drill bit is designed to be used in a power drill motor. The typical auger drill bit is fitted with a screw-like point designed to pull the bit into the timber in a self-feeding manner.

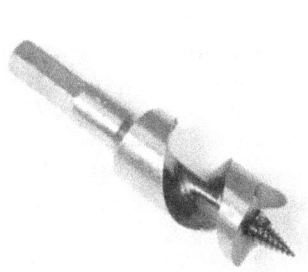

It is easy to distinguish a hand-brace bit from an auger drill bit that is designed to be used in a drill motor. The hand-brace type bit is manufactured with a tapered, square area at the end of the shank designed to lock into the jaws of the hand brace chuck. There are two distinct versions of the auger drill bit: the Jennings and the Irwin. Both bits are similarly designed, however, the Jennings-style auger bit has two flutes that carry the chips up and out of the hole. The Irwin has one flute that goes completely up the drill bit and one that only goes half-way up the bit.

Occasionally called a ribbon bit due to the design of the bit resembling a coiled piece of ribbon, the edges of the wide flutes are typically sharp and can cause injury if not properly used. The screw point is very important, as it holds the bit square in the drilling location and prevents the auger drill bit from wandering when the hole is initially started.

The cutting head of the auger drill bit is virtually two plane-type blades mounted under the centre screw point. Two spurs on the outer edge of the cutting planes are designed to cut the wood fibres as the screw point pulls the bit into the timber. The purpose of the screw point is to pull the auger drill bit into the timber, thereby eliminating the need for the user to force the bit into the timber. The user need only provide the power to turn the drill bit, and this is often seen as a great advantage when using a hand brace.

3.3.2.2 Brad Point Bits:

A brad drill bit, otherwise called a brad-point drill bit, is a tool that is used in woodworking for boring holes that have to have precision circumferences and is most often seen in setting dowel pins but can be used for many applications. Brad drill bits are used with plastics and timber materials that are usually prone to cracking or breaking when a standard drill bit is used.

The difference between a brad drill bit and a standard drill bit can be seen by looking at the cutting tip of the bit. A standard drill bit has a V-shaped tip that terminates in a point at the very bottom. By comparison, a brad drill bit resembles an "M."

The centre point is used to align the drill bit onto the measured point where the hole is to be drilled, and the two edges of the drill bit cut into the wood or plastic in the same manner that a hole saw does. The tips cut the timber or plastic while they spin, so the materials are less likely to splinter or otherwise become damaged. In addition, brad drills will produce smoother side walls than ordinary drill bits.

Using a brad drill bit is similar to using a standard drill bit. A user should mark the point where the hole is to be drilled, then line up the centre point of the drill over this point. The centre point of the brad drill bit should be exactly in the centre of the mark. The power to the drill is then turned on, and the weight of the drill is allowed to bore the hole into the timber. Attempting to press down too hard could damage the timber and defeat the purpose of using the brad drill in the first place.

Brad drill bits are available in as many sizes as it is possible to find standard drill bits in, but consumers should anticipate paying slightly more for brad drill bits because they are more complicated to manufacture. In addition, brad drill bits cannot be sharpened in the same way that standard drill bits can be; this is why they are predominantly used for timber and plastics. Hardwood will tend to heat up the drill bit, which causes it to lose its sharp edge, so the brad drill bit should not be allowed to excessively overheat while it is being used. Correct usage will prolong the tool's life and ensure accurate cuts in all of the user's woodworking projects.

3.2.2.3 Hole Saw Bit:

A hole saw bit is similar to a circular-shaped saw. It is attached to a drill, and used to cut large holes in thin pieces of material, like timber, plastic, soft plaster, or soft metal. For example, a hole saw could be used to make a hole in a door for a doorknob.

A hole saw cuts using sharpened saw teeth which are moulded onto the end of the saw's cylinder that does not attach to a drill. The placement of the saw teeth affects the size of the cut hole. Typically, the hole that a hole saw will cut is slightly larger than the hole saw's cylinder.

An adjustable hole saw can make holes of different sizes. It has a flat metal disc with a series of grooves that make progressively wider circles. The saw teeth can be snapped into the grooves that correspond to the size of the planned hole.

Although holes can be more efficiently cut with a boring bit, boring bits can be expensive and difficult to find. Hole saws can be purchased for as little as a few dollars, work with standard drills, and are easily found at most hardware stores.

| *Fixed Diameter* | *Variable Diameter* | *Variable Diameter* |

It takes a little bit of skill and practice to use this kind of saw to cut a hole without ragged edges. First of all, the operator should make sure that the hole saw is of the proper size. It is also important that the saw is tightly attached to the drill. To make a splinter-free hole, the operator should drill from each side of the material. To ensure a straight hole, they should make sure that the saw teeth are evenly contacted with the material before starting to drill.

Anyone using a hole saw should take care to make sure that it is used safely. If the saw hits a nail while drilling, or if the drill is accidentally tilted, it can forcefully twist; the sudden twist could cause injury to the operator's wrist. It is difficult to predict a hidden nail, so there are ways to guard against this twisting motion. If using a hand powered drill, an operator can brace the drill handle against their hip or leg while drilling, hold the drill with two hands, and use the drill's side handle.

There are several other ways to ensure safe drilling. Operators should not wear loose clothing or jewellery while operating the drill. If the drill has a speed lock button, the user should take care not to press it accidentally while drilling.

3.2.2.4 Masonry Drill Bits:

Masonry drill bits are tools used to drill through masonry structures, including brick and concrete. The drills resemble standard drill bits, but feature a more heavy-duty construction and special materials that allow them to drill through harder surfaces.

Masonry drill bits may be used in conjunction with either a standard drill or a hammer drill, which hammers into the brick or concrete as it rotates. Using a variety of masonry drill bits, contractors can install electrical wiring and pipes through masonry walls, or even attach objects to masonry structures.

Manufacturers typically construct masonry drill bits out of very hard, durable metals instead of the traditional steel used for wood drilling. Some feature tungsten or carbide tips fused to a steel shank, while others include a silicon coating to enhance strength. Depending on the application, the entire length of the bit may be built using tungsten or carbide rather than just the tip; these materials allow the bit to hold up against the high temperatures and levels of force commonly found in masonry drilling.

In all but the most basic masonry drilling applications, workers must rely on hammer drills rather than regular drills. Hammer drills feature a heavy weight that applies a tapping force to the bit; the tapping force, combined with the rotation from the motor, helps to more efficiently drive the bit into concrete or stone. Operators should be aware that hammer drills often feature a different design than a regular drill, and typically require special bits.

When choosing masonry drill bits, users must match the diameter of the shank to the size of the drill's chuck. The chuck is located at the front of the drill, and holds the bit in place during operation. Each type of drill may feature special bitting, or keying, which secures the bit. The bitting system on the drill bit must match the system used on the drill itself in order for the two to function together successfully.

Drilling through masonry creates a large amount of dust, which can impact the performance of the drill and bits. Installers should operate the drill at a slow speed to keep the bit from overheating as it goes through the masonry. It is also helpful to remove the bit from the wall frequently to remove dust from the opening. By keeping the bit spinning as it's removed from the wall, users can draw out excess dust to keep the opening as clean as possible.

3.2.2.5 Metal Drill Bits:

A steel drill bit may refer to either a drill bit made of steel or a drill bit designed to drill through steel. The precise meaning of the term depends on the context in which it is used. Steel is the most commonly-used material in drill bits, and bits are available in different grades of steel with cheaper bits typically being made of less-durable material. The term "steel drill bit" may mean a drill bit that is specifically made for drilling through steel. Tempered steel is a very durable material and specialized drill bits are needed to drill through this type of steel successfully.

Most ordinary drill bits are made of steel. Better bits are typically made of higher-carbon steel, and can withstand more use and abuse before either becoming dull or breaking. Bits designed to drill through masonry are typically made of high-quality steel, and are very hard, but still wear out more rapidly than ordinary drill bits, as masonry causes much more wear and tear on a drill bit even when a hammer drill is used. Heat build-up can shorten the working life of any drill bit as excess heat can ruin the temper of the steel and make it much more prone to breaking.

Some types of steel drill bits are designed to drill through very hard tempered steel. Ordinary drill bits are often able to drill through mild steel, although this causes wear and can lead to excessive heat build-up. Normal drill bits typically fail badly when used to attack high-carbon steel, as they are typically not as hard as the material that they are being used to drill and are thus unable to drill through it effectively.

Specialized steel drill bit sets make use of more expensive steel that is exceptionally hard; this steel is able to drill through other steel with relative ease, although heat build-up can still be a problem. Bits used to drill through a great deal of hard steel should be cooled and lubricated with oil in accordance with the recommendations of their manufacturers. Many larger varieties of steel drill bit are designed to be used in tools that automatically apply a stream of oil to cool the bits.

The most modern and durable steel drill bits are made from high speed steel (HSS); this alloy is designed to survive use in tools that operate at high speeds and is more durable even than the high-carbon steel used in normal steel drilling bits. HSS bits also require cooling when used for a protracted period of time.

The metric or imperial size of a drill is stamped into the shank of the drill.

3.2.2.6 SDS Bits:

Special direct system drills are often referred to as SDS drills and are designed for use with several types of drill bits, including hammer drill bits. Most hammer drills operate by moving a chuck, the hollow channel into which a drill bit locks, back and forth to produce a hammer action. An SDS drill employs bits that feature a groove system that allows these chucks to stay in place while still providing drilling and hammering actions. An SDS drill bit is thought to provide more efficient drilling and hammering action than a regular hammer drill bit.

The shank on an SDS drill bit features a set of specially designed grooves to give the bit rotary force. Closed channels along the shank are designed to accommodate roller balls, which allow SDS drill bits to move back and forth, providing hammer action. The bits allow for quick, nearly seamless bit changing; tools and chuck keys are not needed to change the bits.

SDS drills are more powerful; they generally offer higher torque output and operate more efficiently than the average hammer drills on the market. The bits are also safer because their stationary chuck design reduces slips. A stationary chuck allows the more accurate direction of power from drill to workpiece. An SDS hammer drill system can drill through hard materials such as concrete and brick much more quickly than the average hammer drill system. Despite the increase in performance, these drills usually operate much quieter than regular hammer drills.

SDS drill bits are designed for hammer drilling applications on harder materials such as brick, granite and marble and have to be durable enough to withstand hammering applications while drilling, which is why most manufacturers construct their SDS drill bit models out of tough, durable materials such as tungsten carbide.

Material hardness can be measured using the Moh's scale, which offers values from one to ten. Diamonds are the hardest material on Earth, they offer a value of ten on this scale. Tungsten carbide is also a very hard material with a hardness value of around nine. An SDS drill bit formed out of tungsten carbide can easily drill through and chip away materials such as brick, concrete, and granite, all of which offer lower hardness values than tungsten carbide.

Drilling and hammering through concrete and other hard materials produces a large amount of dust and debris. Wearing a dust mask or respirator system helps SDS drill operators keep potentially harmful airborne debris from entering their lungs while drilling. The loose chips and debris produced by drilling through hard materials can also damage the eyes. SDS drill operators should wear goggles, masks, or other safety equipment to prevent eye damage while drilling with an SDS drill.

3.2.2.7 Tile & Glass Bits:

Tile and glass drill bits are typically used with an electric drill to create holes in ceramic tiles and glass. The bit is coated with diamond pieces that act like sandpaper to rub holes in the material being drilled without cracking or breakage.

One of the main differences between a glass drill bit and one designed for timber or metal is that the glass bit is smooth, straight and has a rough texture; operators familiar with the more common twist bit are used to seeing the spiralling flutes of the bit that both cut into the material and guide the drilled-out matter up and out of the hole. The glass drill bit doesn't have any flutes, because when a hole is drilled in glass the excess material is fine enough that it turns to dust rather than shavings that must be removed from the hole.

Since the bit is coated with diamond chips or diamond dust, much of the bit has the look and feel of sandpaper. During the process of drilling the diamond coating rubs away the glass, creating a hole in the material being drilled. Like sandpaper, the coating comes in different textures, from rough to very fine, so that the coating can be matched to the job.

The bits have some key differences that distinguish them from other types of drill bits. In the glass drill bit, the diamond coating wears off of the outside of the bit over time, eventually rendering it useless. The bit's longevity depends on how hard the surface being drilled is, how thick, and, in some cases, the skill of the operator. The coating may last for many, many uses or it may only last long enough to drill a single hole; it is important to monitor the bit to make sure it is not worn out, since overusing one can lead to cracking and breakage.

A glass drill bit must be kept well lubricated for the best possible results which are most commonly done by running water over the drilling surface. Caution must be used as this practice can be hazardous with electric drills. It is important to avoid getting water on the drill itself while it is connected to a power source. Failure to do so could result in harm to the operator.

3.2.2.8 Wood Spade (or Paddle) Bits:

A spade bit is a type of drill bit used for boring large holes into timber objects; these bits consist of a long, thin rod known as a shank. One end of the shank is inserted into a drill during use. The other end is flat, with a small point projecting out at the end. Between this point are two cutting edges, which are used to cut a cylindrical opening in timber as the bit rotates in the drill.

The bits are available in a wide variety of sizes, and can be used to drill a range of hole sizes. Depending on the quality of the bit being used, it may be necessary to drill a pilot hole in the timber before using the spade bit; this hole is created using a traditional spiral drill bit, and is placed directly in the centre of the planned hole. The pilot hole helps direct the pointed tip of the spade bit, holding it steady to increase the accuracy of a cut.

Some modern spade bits have a threaded tip at the end, similar to a screw. These units are known as self-feeding bits, and often eliminate the need for pilot holes. A self-feeding bit will generally be more expensive than a regular spade bit, but will greatly improve the speed and accuracy of the cuts.

Several other types of tools can be used as an alternative to these spade bits. A hole saw is the most common, and is used to make fast, clean cuts. Auger bits can also be used to cut holes. While they are slower than spade bits, auger bits are also more accurate and precise, though they may not work on harder timber species. To cut very large holes in timber, such as those required for pipes, a multi-spur bit can be used.

One of the primary benefits to spade bits is that they can be easily filed down to create custom-sized holes. Because they are so affordable, most users will simply use a metal file to sand the edges of the bit down to the desired size. The bits are also easy for inexperienced users, and are one of the most basic tools for cutting holes in timber doors, furnishings, and many other applications.

3.3.3 Reamers:

A reamer is a precision cutting tool designed to finish a hole to a specific dia. Since drill bits produce slightly oversized holes, reamers are used where precision tolerances of 0.025 mm are required. Reamers have little if no cutting action on their ends, so a pilot hole is required in preparation to reaming. Some general guidelines for using reamers are:

- The cutting speed for reaming should be about 1/3 of the speed used for drilling operation of the same material.
- Before reaming, leave about 0.25 mm of material on holes up to 12 mm, and about 0.5 mm of material on larger holes.
- Never rotate a reamer in the reverse direction.
- Use the correct cutting fluid for the material.
- Remove the reamer from the hole occasionally while cutting to clear chips, which can cause galling on the surface of the precision hole.
- Never stop the machine with the reamer in the hole.
- Clean and return the reamer to its proper storage place.

3.3.4 Arbor:

A drill arbor is a small spindle-shaped shank found on a drill. It can be a separate small part that is removable from the drill itself or it can be manufactured so that it is already on the end of a drill spindle. An arbor is removed by placing a chuck into a vise and can be unscrewed. In some drills the drill chuck is kept mounted directly onto the arbor during use.

Figure 3.11

3.3.5 Countersink Bit:

Countersinking is an operation in which a cone-shaped enlargement is cut at the top of a hole to form a recess below the surface. A conical cutting tool is used to produce this chamfer. When countersinking, the cutter must be properly aligned with the existing hole, and should be rotated about 1/3 the cutting speed of the drilling operation for the hole. Countersinking is useful in removing burrs from edges of holes, as well as providing a flush fit for flat-headed fasteners.

Figure 3.12

3.3.6 Counterbore Bit:

Counterboring is the process of cylindrically enlarging a hole part way along its length. A counterbore cutter is similar to a drill bit in that it has a shank and fluted body, but instead of a point, it has a smaller diameter pilot portion. The pilot fits into a pre-drilled hole, and guides the counterbore. Therefore the counterbore must be aligned with the original hole, so the pilot will follow the hole properly. Counterbores are used to accommodate studs, bolts, or socket head capscrews where a flush surface application is required.

Large diameter counterbore bits are used to spot-face rough or cast surfaces to provide an even surface for fastenings.

Figure 3.13

3.3.7 Tap:

A tap (Figure 3.14) is a tool used to cut internal threads in a cylindrical hole. A tap is fluted like a drill, but the flutes actually perform the cutting operation. The flutes extend the length of the threaded section and also serve to remove the chips being produced. The tap is fitted into chuck adaptor (Figure 3.15) The most common taps used are:

Figure 3.14

- The starting or tapered tap. This tap is used to start threads. At least the first six threads of this tap are tapered before the full diameter of the thread is reached.
- The plug tap. This is the general use tap, and is used to cut threads after the taper tap has been used and removed. Three to five of its first threads are tapered. This is the last tap used if the hole extends all the way through the workpiece.

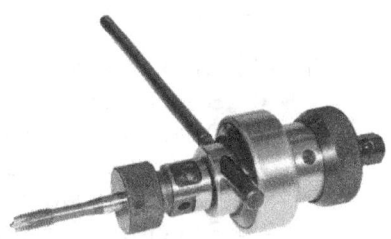

Figure 3.15

Cutting fluids should always be used when tapping holes. It is also recommended to advance the tap one full turn and the reverse it 1/4 turn to break the chip being formed. Always use a tap handle, not pliers or a crescent wrench to turn the tap; they can damage the tap, and the unequal torque provided can cause a thread to be cut poorly.

3.4 Operation of the Drill Press:

Typical sizes range from 500 watts to 2500 watts and a common machine shop drill press can drill holes in mild steel up to 25 mm in diameter. The majority of drill bits are made of high speed steel. Drill bits that are tapered, fit directly in the spindle of the drill press. Regular non-tapered bits, sit into a keyed or keyless chuck. The chuck is then placed into the spindle. Drill presses are typically powered by an electric motor which is mounted at the rear of the head assembly. Power is usually transmitted from the motor to the spindle via V-groove belts or a gear box. Spindle speeds are changed by locating the V-belt on different grooves atop the spindle and motor drive pulleys, or by using handles to change the gear ratio on the gear box. The spindle is forced down using a handle located on the right side of the head assembly. Pulling the handle forward and down moves the spindle down and releasing the handle returns the spindle by spring action; this handle should be pulled with slow steady pressure allowing for occasional releases in pressure to reduce heat and prevent long strands of waste material from interfering or injuring the operator. Both the head assembly and table can usually be adjusted up and down. Some tables allow lifting action to suit the job at hand.

The vice can be secured in two ways, the first way is to directly bolt the vice and the drill press table. The second would be to brace the vice on the left hand side with a lateral support. As the drill rotates clockwise, the material tends to rotate clockwise as well; by placing the material on the operators' left hand side, the force of the rotation will hold the material in place. After the material is firmly secured for drilling, the speed of the press rotation can then be determined; this aspect depends on the hole size and material that is going to be drilled. When selecting the type of drill to be used drill speed, material hardness and size of hole must be considered. In order to drill a hole in a piece of material with a standard drill bit the press must be rotating clockwise. When operating a drill press the operator must always wear safety glasses and never wear loose fitting clothing or gloves; these items could catch in the chuck while it is moving and pull operators into it. Work boots are recommended for use with the press. The boots will help to protect the feet from sharp chips which fall on the floor during drilling, and from pieces of material which may accidently fall off of the drill press table during setup.

3.4.1 Checklist for Drill Press Operation:

Successful operation of the drill press requires the operator to be familiar with the machine and the desired operation.

When drilling large diameter holes through hard materials (e.g. Mild steel), a small diameter pilot (or centre) hole must be first drilled. The centre drill should leave an opening big enough to give the next size diameter drill a place to start cutting. As a rule, the first hole should be drilled using a 3 mm diameter drill bit; subsequent holes are drilled by increasing the diameter by 3 mm until the required drill size is approached. If a hole must be an exact size at finish a reamer must be used.

The following are the recommendations to follow when drilling a hole:
- Select material as determined by design constraints.
- Determine the logical order of fabrication.
- Select requirements for machine tool usage in the fabrication process.
- Determine work to be done by the drill press.
- Select the drill press to be used.
- Center punch the material where it is to be drilled.
- Determine the method of securing the material to the drill press.
- Select the drill bit to be used (tapered, carbide, etc.)
- Secure the drill bit in the chuck or directly in the spindle if it is a tapered bit.
- Determine the proper rotation speed for the conditions.
- Adjust the position of the V-belt to give the selected rpm.
- Check the setup for proper alignment and security.
- Ensure the bit is not touching the material.
- Check to ensure there is no loose fitting clothing.
- Ensure safety glasses are on properly.
- Turn on the Drill Press.
- Using the spindle feed handle, slowly move the bit down to drill the material.
- Move the handle up and down, to reduce heat and break off long strands of waste material.
- Use lubrication as required.
- Shut off drill press to check progress of work.
- When complete, ensure material is not too hot, and remove material.
- Remove drill bit and return to storage.
- Clean up swarf and waste from the floor and machine.

3.5 Drill Press Maintenance:

Similar to all machinery, it is very important to compile a regular maintenance routine for pedestal drills; regular maintenance will increase machine efficiency, productivity & profitability. Geared head drill presses, pedestal drills, radial arm drills are designed for varied industrial tasks and in order to maintain healthy production, it is very essential to maintain the equipment in a safe condition. Regular maintenance will extend the machines efficiency and their life-span. Moreover, it will not cause disruption in the routine activities and is most essential in considering the safety measures for better output.

3.6 Drill Press Safety:

The drill press can be a safe machine, but only as long as the operator is aware of the hazards involved. Chips are produced in great quantities, and must be safety handled. The rotating chuck and cutter can also be a hazard. Develop safe working habits in the use of protective clothes, set-ups, and tools. The following rules must be observed when working on any drill press in the workshop:

1. All work shall be secured using either clamps or a vise to the drill press table. It is unsafe to use the hands to hold any workpiece being drilled.

2. Drill press head and table shall be securely locked to the column before operating the drill press and must always be checked prior to starting the machine.

3. Plan out the work process thoroughly before starting.

4. Know the location of the ON/OFF button.

5. Always use the correct tooling. Tooling shall always be maintained and properly sharpened. All tooling must be run at the recommended speeds and feeds as they apply to the job. Use only recommended accessories and follow the manufacturer's instructions.

6. Tooling shall be not be forced in to any workpiece but fed according to the proper specifications. Failure to follow the instructions will not only ruin the tooling as well as the machine, but can cause serious injury.

7. Never brush away any swarf and chips while the machine is in operation. All clean-up must be carried out when the machine is stopped.

8. Keep hands in sight. Do not put hands or fingers around, on, or below any rotating cutting tools. Leather safety gloves should be used when handling any sharp objects or cutting tools.

9. Always remove the chuck key immediately after using it. A key left in the chuck will be thrown out at a high velocity when the machine is turned on. Never let the chuck key leave your hand except to put it back into its holder.

10. To prevent serious injury, never stop a drill press spindle using the hand after the drill has been turned OFF.

11. Always wear protective eye wear when operating, servicing or adjusting machinery.

12. When drilling in material which causes dust, a dust mask shall be worn.

13. Avoid contact with coolant, especially guarding the eyes.

14. Non-slip footwear and safety shoes are recommended.

15. Wear ear protectors (plugs or muffs) during extended periods of operation.

The most important part of drill press safety, as with all machine tools, is common sense. If the actions that you are considering appear dangerous they probably are. Stop! Consider alternate methods of fabrication, and most of all seek assistance of a supervisor.

3.5.1 Personal Protection Equipment:

 Approved safety glasses must be worn in all workshops.

 Long and loose hair must be contained at all times in all workshops.

 Approved footwear with substantial uppers must be worn in all workshops.

 Close fitting clothing or overalls must be worn in all workshops.

 Exposed rings and jewellery must be removed.

Skill Practice Exercises:

Skill Practice Exercise MEM05005-RQ-0301
Answer the following questions:

1. Identify the following various parts of a Drill Press:

A. _____

B. _____

C. _____

D. _____

E. _____

F. _____

G. _____

H. _____

J. _____

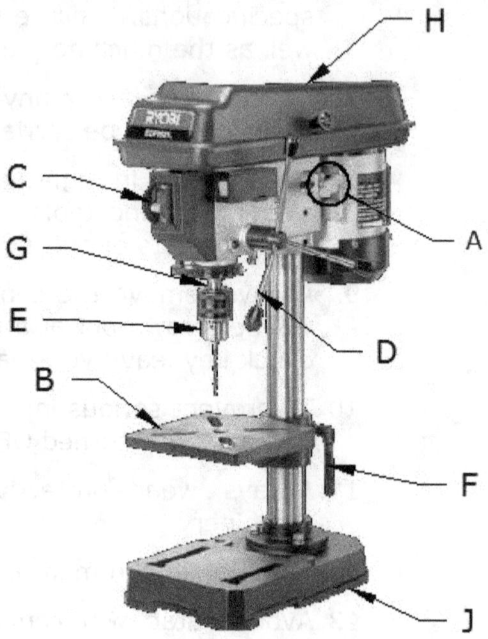

2. A workpiece is secured to:

    A – Pedestal Base

    B – Support Post

    C – Vice

    D - Table

3. Maintain constant awareness of activity in surrounding area.

    TRUE   FALSE

4. Which of the following processes cannot be performed on a drill press?

    A – Drill

    B – Chamfer

    C – Counterbore

    D - Countersink

5. What drilling process would be used to prepare a hole to suit a socket head cap screw?

    A – Counterbore

        B – Drill

        C – Countersink

        D - Mortise

6. Select the correct drill process to prepare a rough surface to permit a nut or bolt head to bear uniformly (flat).

        A – Spot surface

        B – Drill

        C – Countersink

        D – Grinding

7. Which drill bit would be used to drill a hole in bricks?

        A – High Speed

        B – Auger

        C – Brad Point

        D – Masonry

8. It is permissible to leave the chuck key in the chuck while drilling a hole.

        TRUE   FALSE

9. After a hole is drilled on a press drill, the chuck is stopped by:

        A – Using the hands to speed up the work rate.

        B – Use the brake lever.

        C – Allowing the chuck to stop naturally.

        D – Hit the emergency stop switch.

10. The size of a high speed drill bit is recognised by:

        A – Numbers on the end of the shank.

        B – Comparing the bit to a chart.

        C – The size of the fastening.

        D – Associating the drill to an existing hole.

11. Which drilling machine would be used to drill a series of multiple size holes?

        A – Micro Drill

        B – Turret Drill

        C – Radial Arm Drill

        D – Upright Sensitive Drill

12. What is the name of the tool used to remove the drill chuck from the spindle of the drill press?

        A – Chuck Key

        B – Bit Chisel

        C – Drift Pin

        D – Open Spanner

13. When drilling a Ø15 mm hole through steel plate:

        A – The hole is drilled in one motion.

B – A block is placed under the material.

C – The centre is marked with a punch.

D – The surface must be free of rust.

14. Which combination of the following PPE is suitable to be worn when drilling holes through metal?

15. Identify the following drill bits.

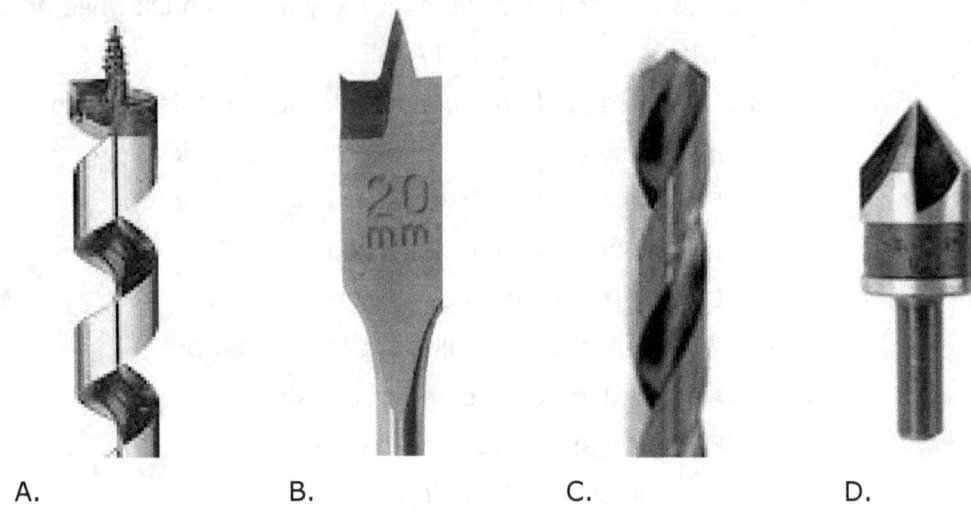

A.   B.   C.   D.

Skill Practice Exercise MEM05005-SP-0302

Use a Drill Press to drill a Ø12 mm hole with a Ø20 mm X 3 mm deep counterbore through 12 mm thick mild steel plate.

Topic 4 – Metal Shears:

Required Skills:
- Identify the different types of metal shears used cutting metal to shape.
- Select a suitable metal shear to suit a specific task.
- Cut various metals into designated lengths and shapes for fabrication.
- Select the correct Personal Protective Equipment to use with individual pieces of machinery.

Required Knowledge:
- Reading and interpreting instructions.
- Use of various metal shears.
- Workplace Health and Safety Regulations.

4.1 Introduction:
Shearing is a simple process whereby a sheet of metal is cut into smaller pieces by two knives which are positioned at an angle relative to each other. The lower knife is firmly attached into a pocket in the stationary table, while the upper blade is fixed to the moving ram assembly. The two blades are separated only by a distance measured in thousandths of an inch at the point of cut.

The basic shear frame consists of table assembly which is welded or bolted to side frames, a moving ram assembly which is powered hydraulically or mechanically and a hold-down ram also fixed to the side frames.

Metal shears are a series of metal working power machines and tools that cut light and heavy metal plate using a "scissor-like" action. Many types of shears used are used to shear sheet metal including guillotines, power shears, bench shears, alligator shears, throatless shears and tin snips.

4.2 Metal Cutting Guillotine:
A guillotine shear is a machine that can shear or cut various materials with a guillotine design. The word "guillotine" is associated with a blade that drops along a vertical track. The guillotine was primarily used in history as a method of execution, particularly in the French Revolution, but the modern guillotine shear cutter is a tool used to form and shape products for a market.

A guillotine shearing machine applies the potential of a dropping blade to an installation for cutting specific kinds of industrial products quickly and precisely. Modern guillotines range in size and cost; some are simple, table-mounted machines while others are bulky, floor-standing installations built to cut a larger piece of metal or other material. Guillotine shearing machines can be operated mechanically or hydraulically.

A guillotine type shear can have a variety of applications in many manufacturing industries. These tools can be used for either wholesale or retail product design; for instance, a sheet metal wholesaler can use a large guillotine machine model to cut simple pieces of raw material for tier pallet shipping. A retail shop can use a different model of guillotine shear to shape metal pieces for specific designs for any kind of retail item.

Figure 4.1

Topic 4 - Metal Shears:

In some metal-working or production shops, guillotine style shears could be part of a tool set that includes new plasma cutting for efficient manufacturing of metal products. Items like plasma cutting and CNC shearing machines represent an increasingly automated system where CAD or Computer Aided Design manages a manufacturing process. A CAD process takes more of the design work off of the human staff and replaces it with neat, capable automated work.

Although more of today's manufacturing process may be automated, guillotine shears and other machines still rely on human staff to keep them operational. Manual machines may still require specific human labour for cutting or shearing. Many of these machine types are inspected by the individual state Workplace Safety Authorities, or other work safety groups. Oversight by authorities and safety groups ensures that potentially dangerous machines like guillotine style cutters are safe for workers to operate in an industrial setting.

Some of the shortcomings of a guillotine shear are that it must run in gibs and ways and therefore need a certain amount of clearance which has a direct effect on the thinnest sheet than can be cut as can be seen in Figure 4.2. The ram moves down with approximately 1° of backward motion which allows the cut sheet to clear the back gauge and drop; sometimes there is not sufficient clearance and the cut part is wedged between the lower blade and the back gauge.

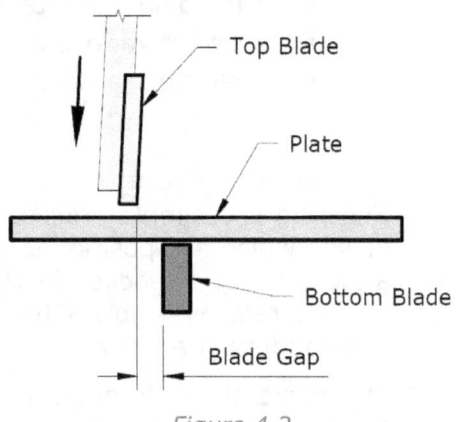

Figure 4.2

When a guillotine shear has a throat it must be heavily re-enforced to avoid the deflection that would normally result from a deep throat.

The apron of the upper ram is heavily gusseted to keep the blades parallel to the bottom blades. This system has worked well for hundreds of years however times change and new engineering becomes available.

A guillotine type shear can have several different applications in the manufacturing industry. They can be used for either wholesale or retail product design i.e. a sheet metal wholesaler can use a large guillotine shear model to cut simple pieces of raw material for tier pallet shipping; a workshop can use a different model of guillotine shear to shape metal pieces for specific designs for any kind of fabrication.

4.2.1 Pre-Operational Safety Check:
- Guillotines are a hazard! Always wear the correct PPE. A hair net needs to be worn when the hair is long and loose clothing or jewellery removed.
- Ensure lighting is turned on and the area both on and around the machine is clean and clear.
- Check the guillotine for any damage, ensuring all guards are fitted correctly.
- Immediately report any damage or malfunctions to the supervisors and maintenance staff.
- Check piece being machined is free from defects such as burrs, kinks or folds.
- Guillotines are for single operator only. Ask for assistance when working with large material.
- Never put hands inside the mesh guarding at the back of the guillotine.

4.2.2 Operating the Guillotine:
- Place metal sheet into the guillotine from the front of the machine only.
- The metal sheet fits under the guard.
- Always keep metal flat against the table.
- Blade gap should be adjusted according to the thickness of the metal plate with no more than 1/30th of the plate thickness.
- Beware of Pinch points. Ensure hands are clear of the yellow clamping guard and cutting blade.

- Slowly push down on foot pedal to engage yellow clamping guard.
- Continue to push down slowly on foot pedal to bring the blade in contact with the work piece. Apply pressure to cut. NEVER jump on the foot pedal, as the blade can be easily damaged.
- Release the pressure on foot pedal gently when cut is complete.
- Dispose of any waste off-cuts.
- Always leave the area on and around the equipment clean and tidy.

4.2.3 Personal Protection Equipment:

 Approved safety glasses must be worn in all workshops.

 Long and loose hair must be contained at all times in all workshops.

 Approved footwear with substantial uppers must be worn in all workshops.

 Close fitting clothing or overalls must be worn in all workshops.

 Exposed rings and jewellery must be removed.

 Gloves must not be worn when using a belt sander.

4.3 Swing Beam:

Swing beam is a term used to describe a shear whereby the upper knife bar moves in an arc around a pivot bearing. Normally the upper knife blade is tapered with only two useable edges to avoid interference with the lower blade during the arcing motion. The Swing beam shear is generally a lower profile design and less weight than the guillotine, but also makes for an easier method of blade clearance adjustment via an eccentric in the pivot bearing.

On a swing beam shear the ram moves on bearings so there is no play what so ever. This allows the swing beam shear to be able to cut paper as long as the blades are sharp. The ram moves from a fulcrum point in the rear of the side frames giving the shear a massive amount of plate between it and the cutting point as shown in Figure 4.3 resulting in almost no detectable deflection.

The back gauge is attached to the bottom of the cutting column and moves up as the blade goes down. This means there will never be a possibility for the material to become stuck between the blade and the back gauge.

Rather than gussets on the apron a swing beam shear wraps the entire ram as one solid gusset making it much stronger than a similarly gusseted ram. It can have a deep throat with no possibility of deflection and can cut even the thickest piece of metal with a very low rake angle.

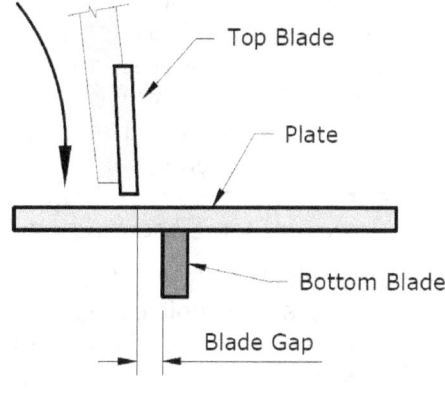

Figure 4.3

4.3.1 Pre-Operational Safety Check:

- Swing beam machines are a hazard! Always wear the correct PPE. A hair net needs to be worn when the hair is long and loose clothing or jewellery removed.
- Ensure lighting is turned on and the area both on and around the machine is clean and clear.
- Check the swing beam for any damage, ensuring all guards are fitted correctly.
- Immediately report any damage or malfunctions to the supervisors and maintenance staff.

Topic 4 - Metal Shears:

- Check piece being machined is free from defects such as burrs, kinks or folds.
- Swing beams are for single operator only. Ask for assistance when working with large material.
- Never put hands inside the mesh guarding at the back of the swing beam.

4.3.2 Operating the Swing Beam:
- Place metal sheet into the swing beam from the front of the machine only.
- The metal sheet fits under the guard.
- Always keep metal flat against the table.
- Blade gap should be adjusted according to the thickness of the metal plate with no more than 1/30th of the plate thickness.
- Beware of Pinch points. Ensure hands are clear of the yellow clamping guard and cutting blade.
- Slowly push down on foot pedal to engage yellow clamping guard.
- Continue to push down slowly on foot pedal to bring the blade in contact with the work piece. Apply pressure to cut. NEVER jump on the foot pedal, as the blade can be easily damaged.
- Release the pressure on foot pedal gently when cut is complete.
- Dispose of any waste off-cuts.
- Always leave the area on and around the equipment clean and tidy.

4.3.3 Personal Protection Equipment:

 Approved safety glasses must be worn in all workshops.

 Long and loose hair must be contained at all times in all workshops.

 Approved footwear with substantial uppers must be worn in all workshops.

 Close fitting clothing or overalls must be worn in all workshops.

 Exposed rings and jewellery must be removed.

 Gloves must not be worn when using a belt sander.

4.4 Bench Shears:

A bench shear, also known as a lever shear, is a bench mounted shear with a compound mechanism to increase the mechanical advantage. The shears are generally used for cutting rough shapes out of medium sized pieces of sheet metal, but cannot do delicate work. For the small shear, it is mostly designed for a wide field of applications; light weight and easy efficient operation, yet very sturdy in construction. The cutting blades fitted are carefully and accurately ground to give easy, clean quick cuts, and free of burrs; these special features help the operators save a great deal of their energy. However, some shearing machines can cut sheet bar and flat bar up to 10 mm.

It is electrically welded together to make it a sturdy stable unit capable to withstand highest stresses due to heavy duty usage. The footplates are reinforced with bracing angles so that they give firm stability to the shear. The machine is provided with section knives with sliding blades which can be adjusted by hand to make 90 cuts on angels and T-sections of different sizes as well as with openings for cutting round and square bars.

Figure 4.4

4.4.1 Pre-Operational Safety Check:
- Locate and ensure you are familiar with all machine operations and controls.
- Ensure all guards are fitted, secure and functional. Do not operate if guards are missing or faulty.
- Ensure bench shears are securely fastened to a bench or purpose-designed stand.
- Ensure shearing edges are in good condition, distortion free and correctly adjusted.
- Ensure working parts are well lubricated and the blades free of rust and dirt.
- Check workspaces and walkways to ensure no slip/trip hazards are present.
- Ensure there is sufficient space around the machine to prevent accidental contact with people in the area..

4.4.2 Operating the Bench Shear:
- Never use bench shears for cutting metal that is beyond the machine's capacity with respect to thickness, shape, or type.
- Material should be properly supported during cutting and industrial type gloves should be worn to protect the hands.
- Use supports for long material and use signage to indicate any tripping hazard.
- Manual handling tasks should be assessed and appropriate procedures put in place.
- Hold material securely to prevent it tilting during the cut.
- Ensure fingers and limbs are clear before operating the bench shears.

4.4.3 Personal Protection Equipment:

 Approved safety glasses must be worn in all workshops.

 Long and loose hair must be contained at all times in all workshops.

 Approved footwear with substantial uppers must be worn in all workshops.

 Close fitting clothing or overalls must be worn in all workshops.

 Protective gloves must be worn when using a bench shear.

 Exposed rings and jewellery must be removed.

4.5 Alligator Shears:
Alligator shears are powered machines designed to shear, or cut, metal items using a hinged, moving blade passing vertically close up against a second, static blade. The static blade generally makes up or is attached to a flat stage on which the metal object lies when being cut. The alligator shears are typically powered by a hydraulic ram that exercises a single cut per cycle. Older models were, however, driven by electric motors and belts, meaning that the shears operated continuously while the motor was activated. Alligator shears are commonly used in metal fabrication and scrap metal facilities to cut a range of metal items including pipes, bars, and even profile steel such as I-beams.

Cutting metal items during manufacturing operations or the preparation of scrap is generally achieved using an oxy-acetylene torch or an alligator shears. Gas cutting is usually employed on very large or tempered items and the shears used for smaller and softer items. The alligator shears have been a standard fixture in this type of installation for many years and offer a quick and effective metal cutting solution.

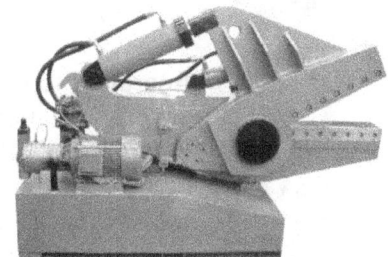

Figure 4.5

Topic 4 - Metal Shears:

Consisting of a hinged upper blade that moves vertically past a static lower blade, the machine certainly resembles its reptilian namesake. The lower blade is generally part of a large, flat table onto which the metal work piece is laid prior to cutting. When the machine is activated, the upper blade moves down, exercising a scissor-like cutting action. The workpiece is then advanced for each consecutive cut; in most cases, a guard descends to enclose the cutting area, preventing accidental amputation or injuries due to work pieces being flung off of the table.

In modern versions of the machine the upper blade is driven by a powerful hydraulic ram that allows for safe, single-cycle operation. Older machines operated with large belt-driven flywheels and ran continuously as long as the drive motor ran; this was not the safest option, as the guard that protected the operator generally lifted to allow for the insertion or movement of the metal being cut. Single-cycle operation allows the alligator's "jaws" to close once only to effect a cut, a far safer option.

These shears are available in a number of sizes, ranging from small, tabletop versions to huge variants capable of exerting several hundred tonnes of cutting pressure. All have operating limitations regarding maximum work piece sizes and the types of material that may be cut though. In general, very hard, tempered materials such as springs and axles should never be cut with an alligator shears as the blades may be damaged or the items could fracture causing serious injuries.

4.5.1 Pre-Operational Safety Check:
- Locate and ensure you are familiar with all machine operations and controls.
- Ensure all guards are fitted, secure and functional. Do not operate if guards are missing or faulty.
- Ensure alligator shears are securely fastened to a bench, purpose-designed stand, or floor.
- Ensure shearing edges are in good condition, distortion free and correctly adjusted.
- Ensure working parts are well lubricated and the blades free of rust and dirt.
- Check workspaces and walkways to ensure no slip/trip hazards are present.
- Ensure there is sufficient space around the machine to prevent accidental contact with people in the area..

4.5.2 Operating the Alligator Shear:
- Never use alligator shears for cutting metal that is beyond the machine's capacity with respect to thickness, shape, or type.
- Material should be properly supported during cutting and industrial type gloves should be worn to protect the hands.
- Use supports for long material and use signage to indicate any tripping hazard.
- Manual handling tasks should be assessed and appropriate procedures put in place.
- Hold material securely to prevent it tilting during the cut.
- Ensure fingers and limbs are clear before operating the alligator shears.

4.5.3 Personal Protection Equipment:

 Approved safety glasses must be worn in all workshops.

 Long and loose hair must be contained at all times in all workshops.

 Approved footwear with substantial uppers must be worn in all workshops.

 Close fitting clothing or overalls must be worn in all workshops.

 Protective gloves must be worn when using a bench shear.

 Exposed rings and jewellery must be removed.

4.6 Power Shears:

A power shear is electrically or pneumatically powered hand tool designed to blank large pieces of sheet metal and are designed to cut straight lines and relatively large radius curves. Power shears are advantageous over a bandsaw because there is not a size limit. Large versions can cut sheet metal up to 2 mm.

An alternative to the hand tools are hydraulically powered tools attached to heavy machinery and are usually used to cut materials in situ that are too bulky to be transported to a cutting facility, too big or dangerous for the hand tools and are stored at remote locations, e.g. mines and remote cattle stations.

Figure 4.6

4.6.1 Pre-Operational Safety Check:
- Inspect the power cord and plug for damage and wear.
- Check the cords are clear of water.
- Locate and ensure you are familiar with all machine operations and controls.
- Ensure shearing edges are in good condition, distortion free and correctly adjusted.
- Ensure working parts are well lubricated and the blades free of rust and dirt.
- Check workspaces and walkways to ensure no slip/trip hazards are present.
- Ensure there is sufficient space around the machine to prevent accidental contact with people in the area.

4.6.2 Operating the Power Shear:
- Never use power shears for cutting metal that is beyond the machine's capacity with respect to thickness, shape, or type.
- Material should be properly supported during cutting and industrial type gloves should be worn to protect the hands.
- Use supports for long material and use signage to indicate any tripping hazard.
- Manual handling tasks should be assessed and appropriate procedures put in place.
- Hold material securely to prevent it tilting during the cut.
- Ensure fingers and limbs are clear before operating the shears.
- Never pull the plug from the socket by "yanking" the cord.

4.6.3 Personal Protection Equipment:

 Approved safety glasses must be worn in all workshops.

 Long and loose hair must be contained at all times in all workshops.

 Approved footwear with substantial uppers must be worn in all workshops.

 Close fitting clothing or overalls must be worn in all workshops.

 Protective gloves must be worn when using power shears.

 Exposed rings and jewellery must be removed.

4.7 Throatless Shears:

Throatless shears are hand operated tools with long blades that are used in metal fabricating, particularly with stainless steel; they are similar to heavy hand shears and to snips, with the exception that throatless shears do not have handles. In addition, throatless shears are designed with just one blade that is anchored to a base. Using a long lever that is attached to the tip of the adjustable blade, the user applies pressure to the metal to be cut.

Throatless shears are a versatile tool in metal fabricating. Using throatless shears, the user can cut sheets of metal to any length desired. In addition, the metal can easily be turned in any direction, since throatless shears do not have handles to get in the way. Consequently, it is possible to follow notches or irregular lines when fabricating with throatless shears. Unlike fabricating with other types of shears, the metal does not become distorted when following unusual cut patterns with throatless shears, because it does not rub against handles or bend as it is forced past them.

Figure 4.7

Many metal fabricators consider throatless shears to be the perfect all-purpose tool for cutting metal because of the flexibility they offer in the types of cuts that can be made. In addition, the same throatless shears can generally be used to cut very heavy gauges of metal, as well as lighter metals, without causing any type of distortion. Throatless shears also have upper and lower blades that are positioned in such a manner that they do not cause knurls, or ridges, to occur, even if used for materials other than metal. In addition, the design of throatless shears ensures that grooves do not appear in the metal being fabricated. The clean cut produced by throatless shears also makes it relatively easy to clean up after completing a job.

Figure 4.8

4.7.1 Pre-Operational Safety Check:
- Check for damaged parts. Before using any tool, any part that appears damaged should be carefully checked to determine that it will operate properly and perform its intended function.
- Check for alignment and binding of all moving parts, broken parts or mounting fixtures and any other condition that may affect proper operation. Any part that is damaged should be properly repaired or replaced by a qualified technician.
- Never use the blades if they are dull, worn, or chipped. Blades must be sharpened, repaired, or replaced before using.
- Check to make sure all bolts are secure but not over-tightened. Check for blade clearance. Make sure clearance is suitable for the workpiece.
- While being used. the lower blade must be tightened. When not being used, the Clamp can be loosened.

4.7.2 Operating the Throatless Shear:
- Adjust the clearance between the Upper Blade and Lower Blade. To decrease the clearance, turn the Adjusting Bolt clockwise. To increase clearance, turn the bolts counter-clockwise. The clearance can be adjusted to between 1/4 and 1/10 of the thickness of the workpiece.
- To gain power and control of the workpiece, start all cuts at the heels of the blades. Circular and irregular cuts are made by turning the workpiece while following through with the downward stroke of the blade.
- Never force the tool or attachment to do the work of a larger industrial tool. It is designed to do the job better and more safely at the rate for which it was intended.

- When a cut has been finished, push the workpiece forward while raising the handle to lift the blade above the workpiece. Start the next cut at the heel of the blade.
- Always keep the material being cut flat on the bed of the shear.

4.7.3 Personal Protection Equipment:

 Approved safety glasses must be worn in all workshops.

 Long and loose hair must be contained at all times in all workshops.

 Approved footwear with substantial uppers must be worn in all workshops.

 Close fitting clothing or overalls must be worn in all workshops.

 Protective gloves must be worn when using throatless shears.

 Exposed rings and jewellery must be removed.

4.8 Tin Snips:

Tin snips are shears designed to cut through thin sheets of sheet metal. While their design resembles scissors, they are much stronger, with heavier blades that make metal cutting as easy as possible. Many hardware stores carry these cutting tools, and there are typically several options to choose from; most workers have several sets in their toolbox for different tasks.

There are three basic types of tin snips: left cutting, straight cutting, and right cutting, and as the names imply the set of the blades on each type cuts in a slightly different way. Straight cutting snips will cut in a straight line, while left and right cutting snips create curved cuts. Because sheet metal is stiff and hard to manoeuvre, it is often necessary to use curved shears to create curved cuts like holes for ductwork.

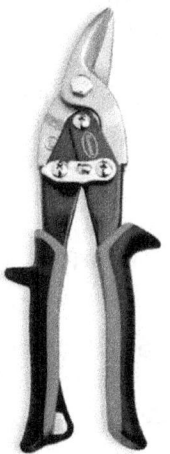  

Figure 4.9 – Left Cut   *Figure 4.10 – Straight Cut*   *Figure 4.11 – Right Cut*

Throatless shears are a versatile tool in metal fabricating. Using throatless shears, the user can cut sheets of metal to any length desired. In addition, the metal can easily be turned in any direction, since throatless shears do not have handles to get in the way. Consequently, it is possible to follow notches or irregular lines when fabricating with throatless shears. Unlike fabricating with other types of shears, the metal does not become distorted when following unusual cut patterns with throatless shears, because it does not rub against handles or bend as it is forced past them.

Most companies colour code their shears to make it easy to know what kind they are. Typically, left cutting snips are red, straight cutting snips are yellow, and right cutting snips are green. The color-coding makes it easy to grab the right pair without inspecting the blade, which can be handy for people in a hurry. As a general rule, these colours are standard across the industry for convenience, although it never hurts to double check when purchasing a set.

The best tin snips have offset handles, which create an angle between the blade and the hands of the user; this means that the user's hands are not caught on the edges of the metal as they cut, making the cutting process faster and safer. Even with offset shears, however, users must wear heavy gloves when cutting sheet metal because the edges can be very sharp, and shards of metal can act like splinters, penetrating the hands and causing pain and discomfort.

Like other bladed tools, tin snips benefit from regular care. The blades should periodically be wiped down and oiled to minimize the risk of rusting. It is also a good idea to sharpen them occasionally to keep their cutting edges crisp, making cutting tasks easier and more comfortable.

4.8.1 Pre-Operational Safety Check:
- Check for damaged parts. Before using any tool, any part that appears damaged should be carefully checked to determine that it will operate properly and perform its intended function.
- Check for alignment and binding of all moving parts, broken parts or mounting fixtures and any other condition that may affect proper operation. Any part that is damaged should be properly repaired or replaced by a qualified technician.
- Never force the tool or attachment to do the work of a larger industrial tool. It is designed to do the job better and more safely at the rate for which it was intended.
- Never use the blades if they are dull, worn, or chipped. Blades must be sharpened, repaired, or replaced before using.

4.8.2 Operating Tin Snips:
- To gain power and control of the workpiece, start all cuts at the heels of the blades.
- Circular and irregular cuts are made by using the left or right cut snips and turning the workpiece while following through with the downward stroke.
- Open the cutters as wide as possible at the start of each stroke, and make long, smooth strokes.
- Lift the cutoff strip and roll it to the side to prevent it from binding on the tin snips' handle
- When a cut has been finished, push the workpiece forward while raising the blades above the workpiece.
- Start the next cut at the heel of the blade.
- Always keep the material being cut flat on the bed of the shear.

4.8.3 Personal Protection Equipment:

 Approved safety glasses must be worn in all workshops.

 Long and loose hair must be contained at all times in all workshops.

 Approved footwear with substantial uppers must be worn in all workshops.

 Close fitting clothing or overalls must be worn in all workshops.

 Protective gloves must be worn when using tin snips.

 Exposed rings and jewellery must be removed.

Skill Practice Exercises:

Skill Practice Exercise MEM05005-RQ-0401

Answer the following questions:

7. What type of pneumatically operated shear would you draw from the store to cut 1.5 mm thick sheetmetal sheets with large curved shapes?

_____

8. Which item of Personal Protection Equipment would not be worn when operating an alligator shear?

      A – Safety Glasses

      B – Close fitting clothing

      C – Breathing Respirator

      D – Protective Gloves

9. When using a guillotine, from where is the material fed into the machine?

_____

10. What is the alternate name for a "Lever Shear"?

_____

11. Which item of Personal Protection Equipment is required to be worn when operating a guillotine?

      A – Loose fitting clothing

      B – Safety Glasses

      C – Breathing Respirator

      D – Protective Gloves

12. Select the correct image that defines a guillotine operation.

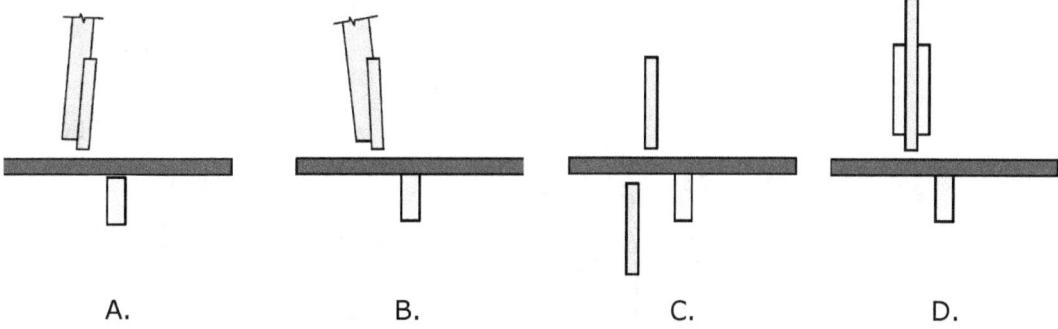

    A.   B.   C.   D.

13. Name the three types of tin snips.

_____

14. Which item of Personal Protection Equipment would not be worn when operating bench shears?

        A – Safety Glasses

        B – Close fitting clothing

        C – Approved Footwear

        D – Protective Gloves

15. Which metal shear cuts metal by the upper knife bar moving in an arc around a pivot bearing?

_____

16. Metal shears work with a:

        A – Rotating action

        B – Reciprocating Action

        C – Scissor action

        D – Buffing action

17. Describe the difference between heavy duty hand shears and throatless shears.

_____

Skill Practice Exercise MEM05005-SP-0402
Cut 10mm mild steel plate into strips 100 mm by 600 mm wide to a tolerance of ±1 mm using metal shears as instructed.

Skill Practice Exercise MEM05005-SP-0403
Cut the following shape from 0.8 mm sheetmetal to a tolerance of ±0.5 mm using tin snips.

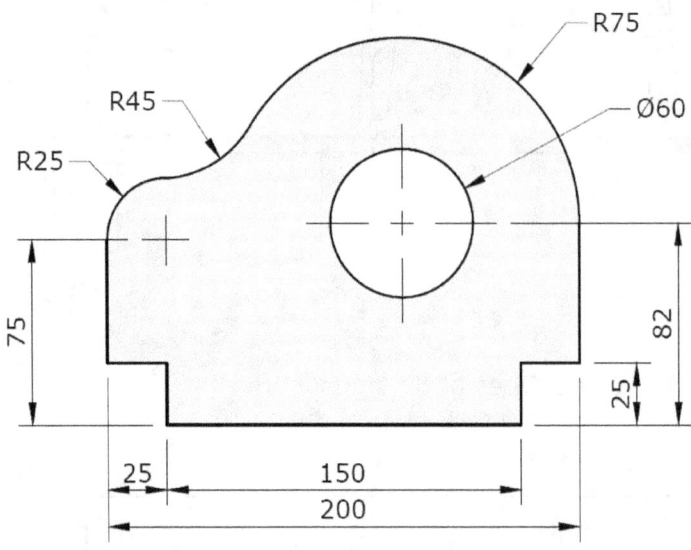

Topic 5 – Cold Saw:

Required Skills:

- Identify the different types of cold saws used in cutting material.
- Select a suitable cold saw to suit a specific task.
- Cut various metals into designated lengths for machining/fabrication/welding.
- Select the correct Personal Protective Equipment to use with individual pieces of machinery.

Required Knowledge:

- Reading and interpreting instructions.
- Use of various cold saws.
- Workplace Health and Safety Regulations.

5.1 Cold Saw:

Cold saws are saws that make use of a circular saw blade to cut through various types of metal, including sheet metal. The name of the saw has to do with the action that takes place during the cutting process, which manages to keep both the metal and the blade from becoming too hot. A cold saw is powered with electricity and is usually a fixed saw in the workshop although portable saws are used on off-workshop sites.

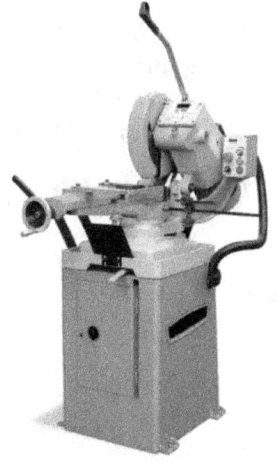

Figure 5.1 – Floor Standing　　　　　*Figure 5.2 - Portable*

The circular saw blades used with a cold saw are often constructed of high speed steel which are resistant to wear even under daily usage. The end result is that it is possible to complete a number of cutting projects before there is a need to replace the blade. High speed steel blades are especially useful when the saws are used for cutting through thicker sections of metal.

Along with the high speed steel blades, a cold saw may also be equipped with a blade that is tipped with tungsten carbide; this type of blade construction also helps to resist wear and tear. One major difference is that tungsten tipped blades can be re-sharpened from time to time, extending the life of the blade; this type of blade is a good fit for use with sheet metal and other metallic components that are relatively thin in design.

One of the advantages of using a cold saw is that the metal and the blade remain relatively cool during the cutting action and is due to the way that the cutting takes place. The design of the blade allows the cutting to focus on the chips created during the action; the chips act as a buffer while not interfering with the cutting action. As a result, the chips tend to collect the heat generated by the cutting and leave both the blade and the metal at a lower temperature.

As part of the support for the cutting to the chips that is created with the configuration of the blade teeth, a cold saw is usually used in conjunction with some type of flood coolant system that allows liquid to run over the blade during use. The action helps to further cool the blade during use and also aids in minimizing dust and sparks during the cutting action.

A portable, cold metal cutting saw easily cuts the thin sheet metal used in the roofing construction which averages approximately 1 mm in thickness and is lightweight, durable and reasonably inexpensive. In addition to sheet metal and corrugated iron, modern metal roofs may be manufactured from stainless steel, copper, aluminum or zinc alloys. A cold metal cutting saw, primarily designed to cut steel sheet metal and corrugated iron, is generally unsuitable for cutting any other material. Electric metal shears or nibblers, metal scissor-type shears or circular saws with abrasive blades would be far more effective in cutting these other types of metal roofing.

5.2 Cold Saw Blades:

High Speed Steel (HSS) cold saw blades come in a variety of diameters and number of teeth, depending on Cold Saw Blades material & type the cold saw blades will be cutting. Rule of thumb is higher number of teeth for thin wall or hard steels, coarse teeth for softer materials and solid materials.

The cold saw blade or wheel can be High Speed Steel (HSS) or Tungsten Carbide Tipped (TCT) depending on material to be cut. The blades are called "cold saw blades" because they transfer all the energy and heat created during the cutting process to the chip which enables the blade and the work material to remain cold. Both types of blades can be resharpened and may be used many times before being discarded. Cold saw blades operate at very slow speeds typically 10-50 RPM when cutting ferrous metal but run faster when cutting nonferrous metal. Cutting speeds and the correct tooth count should be strictly adhered as per the manufacturer's instructions. Typically cold saw blades range 250mm – 400mm, 180 teeth – 400 teeth. Cutting thickness 1.5 to 3.5 mm

Flood coolant or at the least a mist spray is highly recommended for optimal cut quality and blade life. Cold saw blades are commonly listed in millimetres.

The main advantage of a cold saw is there is little heat produced, hence no work hardening in the material. Ideally, they are used for cutting thin wall sections to a close tolerance.

HSS cold saw blades are fed at a constant rate with a very high chip Cold saw Blades. A cold saw cut produces minimal burr, no sparks, no discoloration and no dust.

Figure 5.3

5.3 Changing Saw Blades:

The cold saw must not be connected to the power source when changing saw blades. Failure to comply may result in serious injury! Remove the saw blade as follows

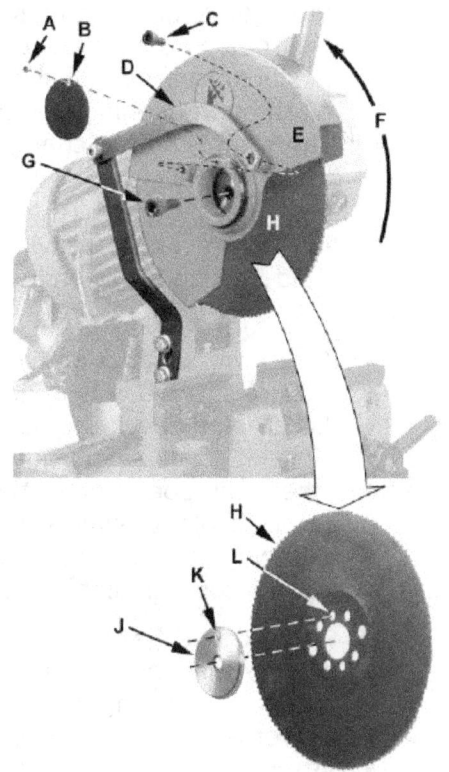

1. Remove the screw (A) and cover plate (B), or loosen the screw just enough to slide the cover plate away to reveal the hub.
2. Remove the socket head cap screw C that secures the bracket (D) to the lower blade guard (E).
3. Raise the lower blade guard (E) upwards (F).
4. Remove socket head cap screw (G) that secures the blade (H). Note: This screw has a left thread requiring a clockwise turn to loosen and a counter clockwise turn to tighten.
5. Remove the blade (H) and outer flange (J). Discard the old blade (or have it resharpened) and replace with a new one.
6. Install the new blade in the reverse order described above. Note that the flange (J) has a pin (K) that needs to match up with the correct keyhole (L), if the blade has more than one keyhole, make the selection such that the mounting holes of the flange and blade are centered.

5.4 Maintenance:

All maintenance must be carried out with the power switched off. Failure to comply may result in serious injury! On completion of maintenance, ensure that replaced parts and/or any tools used have been removed from the machine before starting it up.

5.4.1 Periodic Maintenance:
- Remove chips from the machine and table, preferably with a brush.
- Clean coolant tray and change coolant regularly.
- Top up the coolant level.
- Check the wear of the blade and change if necessary.
- Clean the vise and lubricate all the joints and sliding surfaces using a good quality oil.
- Attend to daily, weekly and annual lubrication recommendations.

5.4.2 Coolant:

This coolant system can operate with either a soluble oil base coolant or water-soluble synthetic coolant. Coolant should be changed regularly.

Coolant Type:
- Soluble Oil Base
- Water-Soluble Synthetic Coolant

The coolant tray is normally located under the cabinet stand. Remove any screws and washers and slide open. Check coolant level in the tray periodically and top off if necessary. Coolant may also be added by pouring directly on the table which will drain into the tank through the chip strainer.

5.4.3 Lubrication:

For long life and trouble free operation, it is essential that this machine is kept clean and well lubricated.

- Vice and leadscrew – oil daily
- Pivot joints and bearings – grease weekly
- Gearbox oil – check level weekly, full level is top of sight glass with head in full up position; gearbox oil should be changed annually. Unscrew operating arm and add oil through opening in the crankcase.
- Lubricant – add or change.

5.5 Cold Saw Operating Procedures:

The following procedure provides detailed instructions for conducting routine operation, on the Cold Saw and is intended as a guide only:

5.5.1 Pre-Operational Safety Checks

- Check workspaces and walkways to ensure that no slip/trip hazards are present.
- Ensure saw blade is in good condition.
- Locate and check the operation of the ON/OFF starter.
- Check that all safety guards are in working order.
- Check the operation of the work vice.
- Check coolant delivery system to allow for sufficient flow of coolant.
- Faulty equipment must not be used and must immediately be reported to the supervisor or maintenance section.

5.5.2 Operational Safety Checks

- Ensure that the work piece is securely held in the work vice.
- Support overhanging work and signpost if it presents a hazard.
- Never leave the machine running unattended.
- Attention must be paid to unusual noises during the sawing process.
- Never force the saw into the workpiece. Use a slow and even feed rate.
- Before making adjustments or before cleaning swarf accumulations switch off and bring the machine to a complete standstill.
- Immediately absorb any coolant spills.
- Switch the machine off after use.
- Leave the machine in a clean, tidy and safe condition.

5.5.3 Personal Protection Equipment:

 Approved safety glasses must be worn in all workshops.

 Long and loose hair must be contained at all times in all workshops.

 Approved footwear with substantial uppers must be worn in all workshops.

 Close fitting clothing or overalls must be worn in all workshops.

 Hearing protection should be worn when noise levels are excessive.

 Exposed rings and jewellery must be removed.

Skill Practice Exercises:

Skill Practice Exercise MEM05005-RQ-0501

Answer the following questions:

1. List a coolant that would be used with a cold saw?

   _____

2. Which item of clothing must be worn when using a cold saw?

   A – Enclosed sports shoes

   B – Spectacles

   C – Hair net (if hair is long)

   D – Dust coat

3. High Speed Steed is used to manufacture\ cold saw blades; in what other material are the blades manufactured?

   _____

4. In preparing to use a cold saw, which of the following must always be done?

   A – Shape the material to the correct length.

   B – Check the saw blade for cracks & missing teeth.

   C – Grind all rust, scale and burrs from the stock.

   D – Remove any guard stopping the sawing action.

5. When using a cold saw, the operator must always:

   A – Remove swarf build-up from the turning blades.

   B – Secure the work piece in the work vice.

   C – Observe 450 volt power is connected to the saw.

   D – Test the guards prevent blockages.

6. Identify the cold sore from the following images.

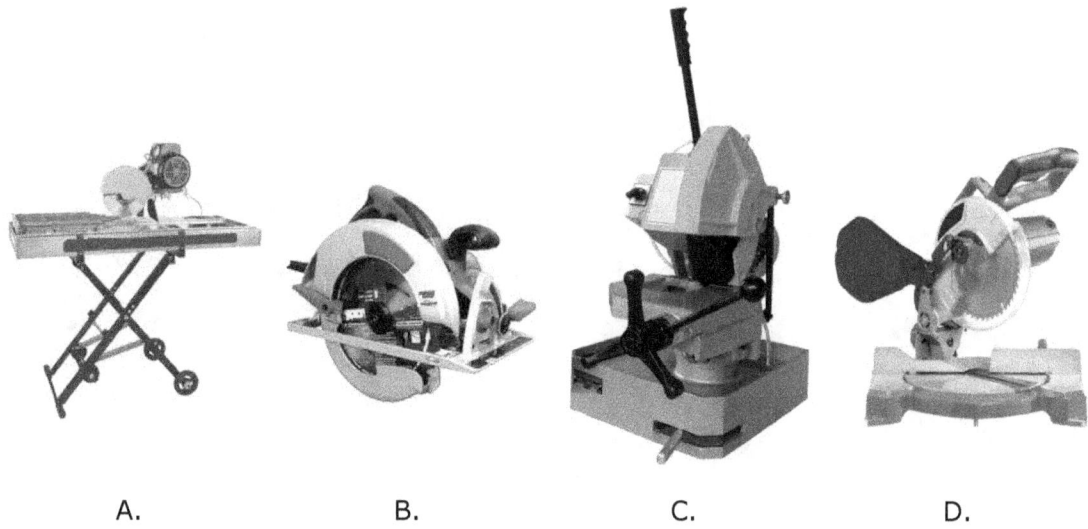

A.   B.   C.   D.

Skill Practice Exercise MEM05005-SP-0502

Cut lengths of mild steel flat bar and structural sections into lengths as instructed by your teacher/supervisor/assessor using a cold saw.

Topic 6 – Band Saw:

Required Skills:
- Identify the different types of band saws used in cutting material.
- Select a suitable band saw to suit a specific task.
- Cut various materials into designated lengths for fabrication/welding.
- Select the correct Personal Protective Equipment to use with individual pieces of machinery.

Required Knowledge:
- Reading and interpreting instructions.
- Use of various band saws.
- Workplace Health and Safety Regulations.

6.1 Band Saw:

Automatic and manual metal cutting band saws are band saws which have been specifically designed for use in cutting metals. While regular band saws can sometimes be equipped with blades which are capable of handling metal, they lack the special features of a true metal cutting band saw, which has been expressly customized for use in working with metals.

Band saws work by running a toothed belt between two wheels. When the band saw is turned on, the belt starts moving, allowing the operator to run objects though the saw for cutting to shape and length. One of the advantages to a band saw is that it cuts very evenly, with a highly regular distribution of pressure.

Band saws for cutting metal are available in vertical and horizontal designs. Typical band speeds range from 12 meters per minute to 1,500 meters per minute; however specialized band saws are built for friction cutting of hard metals and run band are designed to run at speeds to 4,500 meters per minute.

In the case of a metal cutting band saw, the machine usually has a source of lubrication and cooling for the blade which is designed to prevent the blade from overheating while it cuts. The cooling is especially important with thick or hard metals. Water is a simple form of lubricant, although others lubricants are available, with the liquid classically pouring continuously over the blade as the band saw is in operation. The coolant makes the band saw safer and more reliable to operate, and prolongs the life of the blade.

A metal cutting band saw can also be mounted with brushers that continuously clean the blade to remove burrs, chips of metal, and clumps of material. The brushers keep the surface of the blade clean, ensuring a clean cut which will result in crisp edges. Keeping the blade clean also keeps the pressure even, and reduces the risk of accidents, false cuts, and other problems in addition to limiting wear on the blade so that it will last longer.

Both bench-mounted and floor-mounted models of the metal cutting band saw are available, as well as models which can accommodate a range of blade types of and sizes for various applications. Vertical and horizontal metal cutting band saw options are also available. When evaluating band saws for purchase, consumers should think about the type of metal they will be cutting and the kind of cuts they meet to make, and they may want to think about the metal cutting band saw blades they will be using, as they want to be sure that they purchase a saw which will accommodate the blades they want to use.

6.1.1 Horizontal Band Saw:

Horizontal bandsaws hold the workpiece stationary while the blade swings down through the cut. The saws are used to cut long materials such as pipe or bar stock to length; therefore, it is an important part of the facilities in most machine shops. The horizontal design is not useful for cutting curves or complicated shapes. Small horizontal bandsaws typically employ a gravity feed alone, retarded to an adjustable degree by a coil spring; on industrial models, the rate of descent is usually controlled by a hydraulic cylinder bleeding through an adjustable valve. When the saw is set up for a cut, the operator raises the saw, positions the material to be cut underneath the blade, and then turns on the saw. The blade slowly descends into the material, cutting it as the band blade moves. When the cut is complete, a switch is tripped and the saw automatically turns off. More sophisticated versions of this type of saw are partially or entirely automated (via PLC or CNC) for high-volume cutting of machining blanks. Such machines provide a stream of cutting fluid recirculated from a sump, in the same manner that a CNC machining centre does.

6.1.2 Vertical Band Saw:

A vertical band saw, also called a contour saw, keeps the blade's path stationary while the workpiece is moved across it; this type of saw can be used to cut out complex shapes and angles. The part may be fed into the blade manually or with a power assist mechanism. The vertical metal-cutting band saw is often equipped with a built-in blade welder which not only allows the operator to repair broken blades or fabricate new blades quickly, but also allows for the blade to be purposely cut, routed through the centre of a part, and re-welded in order to make interior cuts. These saws are often fitted with a built-in air blower to cool the blade and to blow chips away from the cut area giving the operator a clear view of the work.

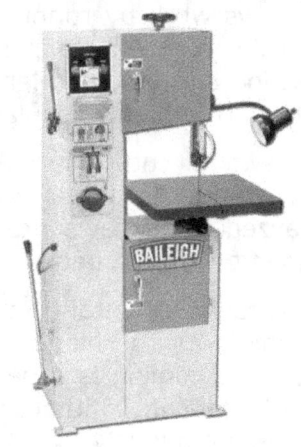

Figure 6.1 – Horizontal Band Saw   Figure 6.2 – Vertical Band Saw

6.2 Metal Cutting Band Saw Blades:

Metal cutting band saw blades use carbon tool steel or bimetal blades. Carbon Tool Steel blades are more economical to purchase. Carbon Tool Steel blades will cut mild steel if used at speeds under 60 meters per minute, preferably with coolant. If sawing in a production setting, the saw is in good repair and adjusted correctly, and want the longest life blade available, then Bimetal bandsaw blades should be used. Bimetal blades cost more than carbon blades, but are generally more economical to operate in the long run, because they can outlast carbon blades by up to 10 times if used properly. Also, they are capable of cutting harder materials, such as stainless steel.

Horizontal metal cutting band saws are typically designed to use only one width of blade. Vertical metal cutting bandsaws have the capacity to run a wide range of widths.

The correct tooth pattern for metal cutting is determined by the thickness (cross-section) of material being cut. At all times there should be between 2 and approximately 10 teeth in contact with the material. In Australia, the pitch of blades is still referred to in Teeth

Per Inch (TPI); generally, fewer teeth per inch are used for thicker materials and more teeth per inch for thinner materials.

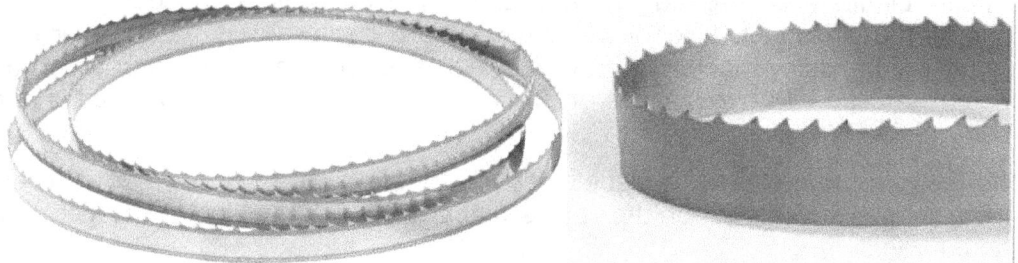

Figure 6.3 – Metal Cutting Band Saw Blades

Advancements have also been made in the band saw blades used to cut metals. The developments of new tooth geometries and tooth pitches have produced increased production rates and greater blade life. New materials and processes such as the cryogenic treatment of blades have produced results that were thought impossible just a few years ago. New machines have been developed to automate the welding process of bandsaw blades as well. Ideal computerized welding machines, setting and cut to length machines and contributions from other manufacturers continue to increase productivity.

An indication of a cracked blade is a rhythmic click as the cracked portion of the blade passes through the material.

6.2.1 Blade Pitch:

Soft or sticky materials and thick stock require coarse tooth blades to provide adequate chip clearance. Hard materials generally require finer tooth blades which are also necessary if a good finish is required. Since two or three teeth of the band saw blade must be in contact with the workpiece at all times to prevent chatter and shearing of teeth, fine tooth blades are used to cut sheet metal and tubing. If sheet metal is too thin to meet this requirement with the finest tooth blade available, the metal should first be mounted on plywood, fibre, or thicker metal to stiffen it. The following table may be used as a guide in selecting the proper pitch bandsaw blade for different metals and metal thicknesses. If the stock is exceptionally large, coarser tooth blades than those recommended for solid stock may be used. **Note**, two or more teeth must contact the workpiece at all times to prevent shearing of the blade teeth. If the recommended pitch for solid stock fails to meet this requirement, a blade with finer pitch must be selected.

	Material	Blade Pitch
Sheetmetal	less than 3 mm thick	24 - 32
	over 3 mm thick	18
Solid Stock		
	Aluminium	6 – 10
	Brass	10 – 12
	Bronze	12 – 14
	Cast Iron	10 – 12
	Copper	10 – 12
	Steel, Alloy	12 – 14
	Steel, High-Speed	12 – 14
	Steel, Machine	10 – 14
	Steel, Stainless	12 – 14
	Steel, Tool	12 – 14
Tubing		
	Less than 3 mm wall thickness	24 – 32
	Over 3 mm wall thickness	18

6.2.2 Blade Width:

When straight sawing, the widest blade available of the proper pitch should be used. Thinner blades are required for contour sawing to prevent the body of the blade from rubbing the sides of the cut when cutting sharp curves. When curves or radii are to be cut on the bandsaw machine, the widest blade adaptable to the sharpest radius to be cut should be used. Narrow blades are more easily broken than wide blades and should be used only where necessary.

The following table lists the blade sizes which can be used for cutting different size radii. If the proper size blade for the radius to be cut is not available, the next size narrower should be used.

Radius to be Cut (mm)	Width of Blade (mm)
3 - 6	3
6 – 8	5
8 – 22	6
22 – 35	8
35 – 62	10
62 and larger	12

6.2.3 Blade Speed:

The cutting speed of a band saw machine is the speed of the band saw blade as it passes the table, measured in metres minute. Proper bandsaw speeds are important in conserving band saw blades. Too great a speed for the material being cut will cause abnormally rapid blade wear. Too slow a speed will result in inefficient production. In general, the harder material requires the slower speed to be selected; conversely, the softer the material, the faster the speed that should be selected. Finer finishes are produced on the cut surface using fasters speeds; the principle applies to light feeds in conjunction with fast speeds. The following table shows the recommended sawing speeds for different materials. In general, the faster speeds should be used to saw thin materials, and the slower speeds should be used for thick materials.

Material	Blade Speed (metres per second)
Aluminium	60 – 600
Brass, Soft	60 – 275
Brass, Hard	20 – 50
Brass, Sheet	60 – 275
Bronze	20 – 50
Cast Iron	15 – 30
Copper	35 – 50
Monel Metal	15 – 30
Rubber, Hard	50 – 75
Steel, Alloy	15 – 30
Steel, High Carbon	15 – 30
Steel, High Speed	15 – 25
Steel, Machine	20 – 50
Steel, Sheet	45 – 60
Steel, Stainless	15 – 20
Steel, Tool	15 - 45

6.3 Arrangement of Material:

It is simpler to determine the correct tooth pattern for solids than for structural shapes. Structural shapes such as square tubing and angle pieces need to be cut at the correct angle to keep the teeth in contact with the thinner portion of the cut. For example, if cutting a piece of 100 x 75 x 8 angle bar, the tooth pattern should be matched to the 8 mm and not to the 100 mm or 75 mm legs. Therefore, the angle bar should be placed in the saw so it sits as a pyramid instead of an "L". The following diagrams show the correct orientation for cutting various structural shapes.

Stack cutting should always be avoided where possible, because regardless how the material is clamped in the vise there will always be varying thicknesses to cut and the teeth cannot be correctly matched. Vibration is also a major problem when stack cutting; when stack cutting, ensure the stack is banded or clamped together as tightly as possible to reduce vibration or movement between the pieces.

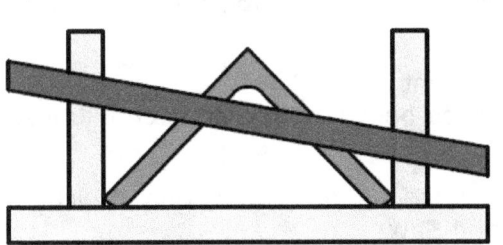 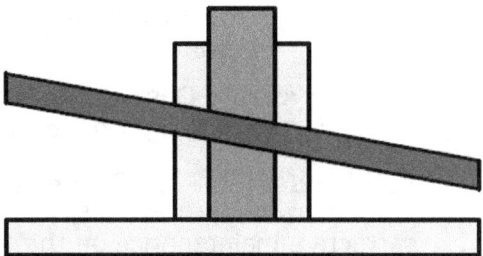

Correct Technique for Cutting Structural Shapes

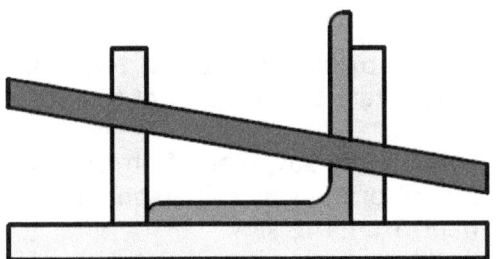 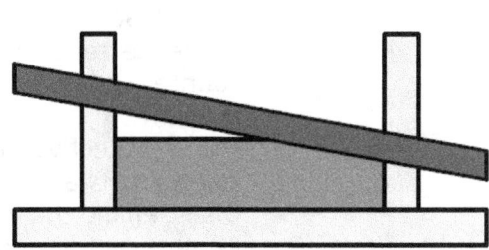

Incorrect Method

6.4 Maintenance:

6.4.1 Preventive Maintenance:

The following steps are used when performing preventive typical maintenance on the Metal Cutting Band Saw:

1. Lubricate Pivot Pin
   - Ensure that power is off to the saw.
   - Open pivot pin oil cups and check oil levels.
   - Add 10 or 20 weight oil to each cup as needed based on current oil levels.

2. Blade Wheel Bearing Greasing (Once per year)
   - Ensure power is off to the saw.
   - Raise blade cover.
   - Decrease tension on the saw blade by turning the blade tension handle counter clockwise (CCW) enough to facilitate the removal of the blade from the saw.
   - Using protective gloves, carefully remove the saw blade from the drive and idler wheels and place it somewhere out of the way of personnel.
   - Disassemble the idler and drive wheels from the saw.
   - Remove idler and drive wheel support bearings from their respective mounting locations.

- Clean and repack bearing with a No.2 general purpose grease. NOTE: If after cleaning the bearings they appear, by feel, to be rough, replace with the appropriate bearing(s).
- Reinstall bearings in their respective mounting locations.
- Reinstall the blade wheels on the saw head.
- Using protective gloves, replace the saw blade on the wheels making sure that the teeth are pointed in the correct direction (teeth pointed in the direction of rotation or movement of the blade)
- Apply enough tension to the saw blade by turning the tension handle in the clockwise (CW) direction to keep it from falling off the blade wheels.
- Turn saw on and tighten blade tension handle (CW) until all blade vibration or flutter disappears.
- Increase tension on the blade 1½ additional turns of the tension handle in the clockwise direction.
- Close the cover and turn the saw off by depressing the "Off" pushbutton switch.

3. Hydraulic Dash-Pot Cylinder Oil Replenishment
   - Add light oil to the cylinder as needed based on operating conditions.

6.4.2 Corrective Maintenance

The following procedures provide detailed instructions for conducting typical routine corrective maintenance on the Metal Cutting Band Saw:

1. Blade Wheel Bearing Replacement
   - Ensure power is off to the saw.
   - Raise blade cover.
   - Decrease tension on the saw blade by turning the blade tension handle counter clockwise (CCW) enough to facilitate the removal of the blade from the saw.
   - Using protective gloves, carefully remove the saw blade from the drive and idler wheels and place it somewhere out of the way of personnel.
   - Disassemble the idler and drive wheels from the saw.
   - Remove idler and drive wheel support bearings from their respective mounting locations.
   - Replace with like bearings.
   - Reinstall bearings in their respective mounting locations.
   - Reinstall the blade wheels on the saw head.
   - Using protective gloves, replace the saw blade on the wheels making sure that the teeth are pointed in the correct direction (teeth pointed in the direction of rotation or movement of the blade)
   - Apply enough tension to the saw blade by turning the tension handle in the clockwise (CW) direction to keep it from falling off the blade wheels.
   - Turn saw on and tighten blade tension handle (CW) until all blade vibration or flutter disappears.
   - Increase tension on the blade 1½ additional turns of the tension handle in the clockwise direction.
   - Close the cover and turn the saw off by depressing the "Off" pushbutton switch.

2. Blade Replacement
   - Ensure power is off to the saw.
   - Raise blade cover.
   - Decrease tension on the saw blade by turning the blade tension handle counter clockwise (CCW) enough to facilitate the removal of the old blade from the saw.
   - Using protective gloves, carefully remove the saw blade from the drive and idler wheels and dispose of it properly.
   - Using protective gloves, carefully uncoil the new replacement blade.

- Place the saw blade on the wheels making sure that the teeth are pointed in the correct direction (teeth pointed in the direction of rotation or movement of the blade)
- Apply enough tension to the saw blade by turning the tension handle in the clockwise (CW) direction to keep it from falling off the blade wheels.
- Turn saw on and tighten blade tension handle (CW) until all blade vibration or flutter disappears.
- Increase tension on the blade 1½ additional turns of the tension handle in the clockwise direction.
- Close the cover and turn the saw off by depressing the "Off" pushbutton switch.

6.4.3 Blade Replacement:

An indication of a cracked blade is a rhythmic click as the cracked portion of the blade passes through the material. The following procedures provide detailed instructions for conducting typical routine saw blade change on the Metal Cutting Band Saw:

- Ensure power is off to the saw.
- Raise blade cover.
- Decrease tension on the saw blade by turning the blade tension handle counter clockwise (CCW) enough to facilitate the removal of the old blade from the saw.
- Using protective gloves, carefully remove the saw blade from the drive and idler wheels and dispose of it properly.
- Using protective gloves, carefully uncoil the new replacement blade.
- Place the saw blade on the wheels making sure that the teeth are pointed in the correct direction (teeth pointed in the direction of rotation or movement of the blade)
- Apply enough tension to the saw blade by turning the tension handle in the clockwise (CW) direction to keep it from falling off the blade wheels.
- Turn saw on and tighten blade tension handle (CW) until all blade vibration or flutter disappears.
- Increase tension on the blade 1½ additional turns of the tension handle in the clockwise direction.
- Close the cover and turn the saw off by depressing the "Off" pushbutton switch.

6.5 Operating a Band Saw:

The following procedure provides detailed instructions for conducting routine operation, on the Metal Cutting Band Saw and is intended as a guide only:

- Check that the saw bed is clear of any debris or previously cut material scraps.
- Raise the saw head to clear the stock (material) being cut.
- Place material supports as need for the length of stock (material) being placed in the saw.
- Place material to be cut between the vise jaws and under the saw blade.
- Bring the vise jaws up to the material but do not tighten until the material cut point is established under the blade.
- Lower the saw head using the hydraulic dash-pot cylinder to a point just above the material to be cut.
- Adjust the blade support guides as needed to clear both sides of the material in the vise, maintaining a distance from 3 mm to 6 mm from the material.
- Measure and adjust the material as needed to attain the desired cut piece length.
- Tighten the vise jaws, start the saw and open the lubricant/cooling fluid supply valve for the desired amount of lubricant/cooling fluid flow. (If not dry cutting.)
- Lower the saw head to the material by slowly opening the needle valve on the hydraulic dash-pot cylinder to start the blade cutting the material.
  Once the blade is cutting the material, open the hydraulic dash-pot cylinder needle valve all the way to allow the saw head pressure to feed the blade through

the material. NOTE: In some cases it will be necessary to use the hydraulic dash-pot cylinder to control the downward feed of the saw through the material.

- Once the saw cuts through the material it will automatically stop when shutoff device depresses the saw "Off" pushbutton switch.
- Shut the lubricant/coolant supply valve.
- Close the hydraulic dash-pot cylinder valve and raise the saw head up to clear it from the work piece. (With the valve closed the saw will remain up away from the bed of the saw)
- Loosen the vise jaws and remove the material/stock from the saw and place it back in the appropriate storage location.
- Remove the cut off piece from the saw.
- Clean up any lubricant/coolant that has trailed its way to floor and wipe any chips or other debris from the saw table and vise jaws.

6.5.1 Personal Protection Equipment:

 Approved safety glasses must be worn in all workshops.

 Long and loose hair must be contained at all times in all workshops.

 Approved footwear with substantial uppers must be worn in all workshops.

 Close fitting clothing or overalls must be worn in all workshops.

 Approved safety glasses must be worn in all workshops.

 Gloves must not be worn when using a belt sander.

 Exposed rings and jewellery must be removed.

Skill Practice Exercises:

Skill Practice Exercise MEM05005-SP-0601

Answer the following questions:

1. The blade guard should be adjusted to what distance above the stock?

    A – 6 mm

    B – 8 mm

    C – 10 mm

    D – 12 mm

2. When sawing small stock which of the following should be used?

    A – Push stick.

    B – Line guard.

    C – Hand rail.

    D – Round guide.

3. While operating a metal band saw, a rhythmic click is heard. What does this indicate?

    A – The motor requires lubrication.

    B – The guard nuts are loose.

    C – The stock is under stress.

    D – The blade is cracked.

4. While operating the band saw, it is a safe practice to?

    A – Open bandsaw covers very carefully.

    B – Keep the work area clear of debris.

    C – Reach across material being cut on table.

    D – Wait until the end of class to clean up scraps.

5. The correct blade width to use for a specific job depends on the:

    A – Cutting speed.

    B – Style of blade.

    C – Radius of the cut.

    D – Type of material.

6. Which PPE must not be worn while operating a band saw?

A.   B.   C.   D.

7. The best time to remove scrap material from the band saw is:

        A – During the cutting.

        B – When the saw stops.

        C – After excessive scrap build-up.

        D – Depends on job.

Skill Practice Exercise MEM05005-SP-0602

Cut lengths of mild steel flat bar and structural sections into lengths as instructed by your teacher/supervisor/assessor using a metal band saw.

Practice Competency Test

Part A:

1. Which of the following lathes would be used to manufacture a brass tapered shaft?

> A – Engine Lathe
> B – Spinning Lathe
> C – Turret Lathe
> D – Wood Lathe

2. Identify the type of drill bit indicated in Image PT1.

3. What is the recommended blade gap on a guillotine?

4. From which two materials are cold saw blades manufactured?

5. Identify the type of machine indicated in Image PT2.

6. If an operator is using a yellow set of tin snips, what type of cut will they produce?

7. What is the maximum speed a band saw works for friction cutting of hard metals?

8. Describe a Radial Arm Drill Press.

9. Identify the type of drill press indicated in Image PT3.

10. What is the normal angle the point of a high speed drill is sharpened?

11. What is used to tighten a drill bit in the drill?

> A – Chuck Key
>
> B – Spindle Lock
>
> C – Drill Spanner
>
> D – Operator's Hand

12. Identify the Swing Beam Metal Shear. (Circle the correct letter)

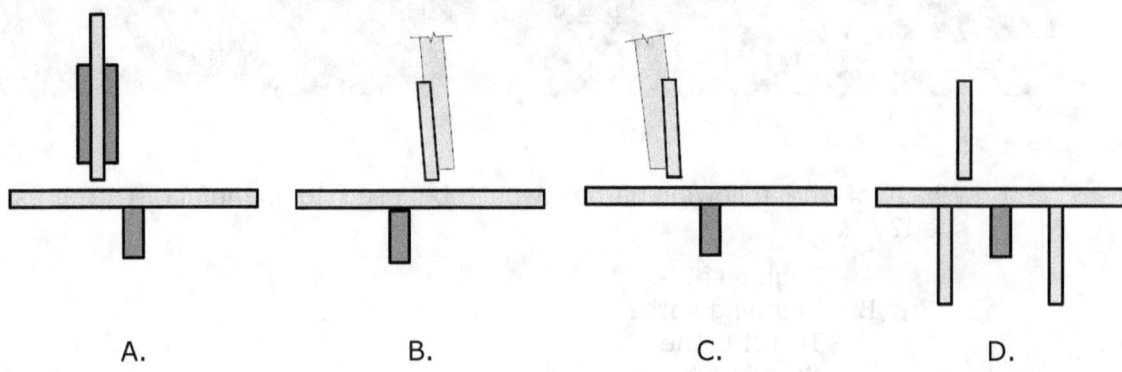

A.   B.   C.   D.

13. What are reamers?

_____

_____

14. Identify the type of tool indicated in Image PT4.

_____

15. What pitch of teeth would be required to cut 3.2 mm sheetmetal?

_____

16. What are 2 main disadvantages for a machine shop purchasing a CNC machine?

_____

17. When are chips and debris removed around the blade on a cold saw?

_____

18. Describe why stack cutting is avoided when cutting structural shapes.

_____

_____

_____

19. Describe the difference between counterbore and countersink bits.

_____

_____

_____

20. What type of machine would the component in Image PT5?

_____

21. If a component to be manufactured required a series of cuts to be made in succession, which metal turning machine would be best suited to the job?

_____

22. What saw blade width would be required to cut a 52 mm radius curve?

_____

23. Identify the type of tool indicated in Image PT6.

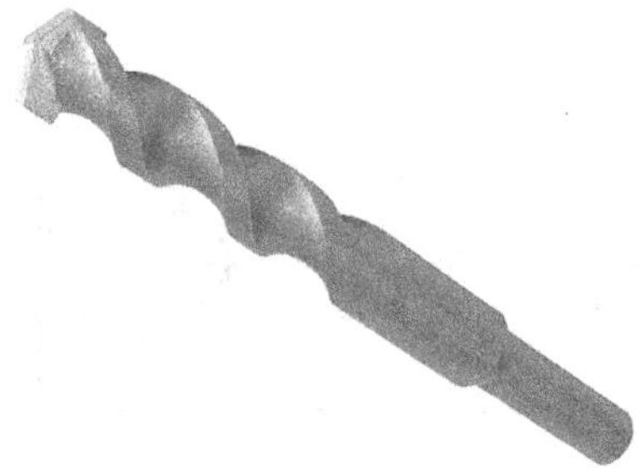

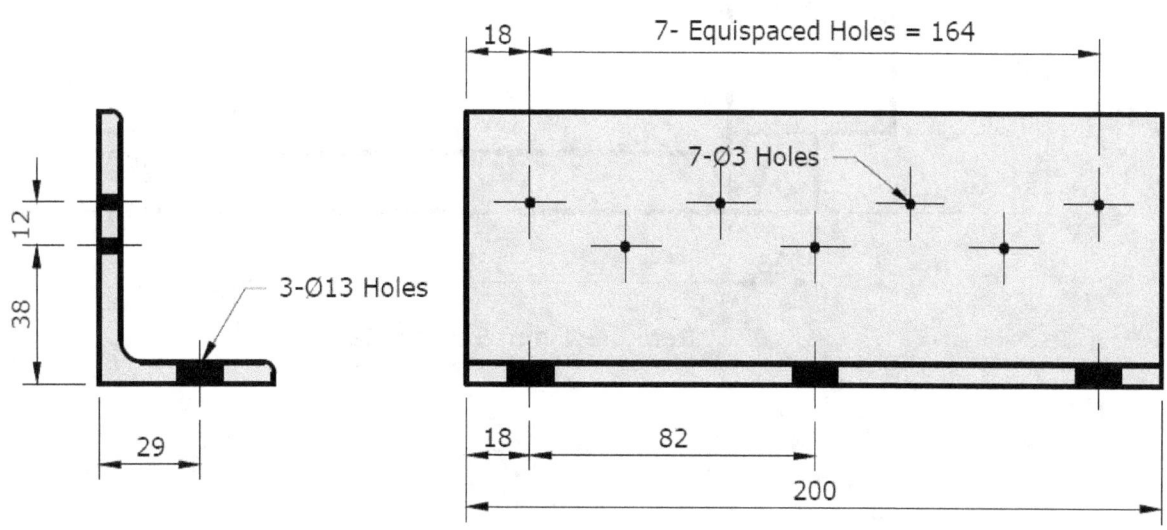

PT1    PT2 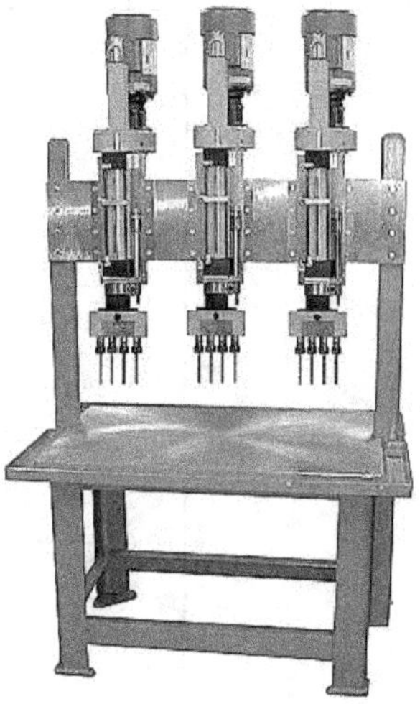   PT3

PT4 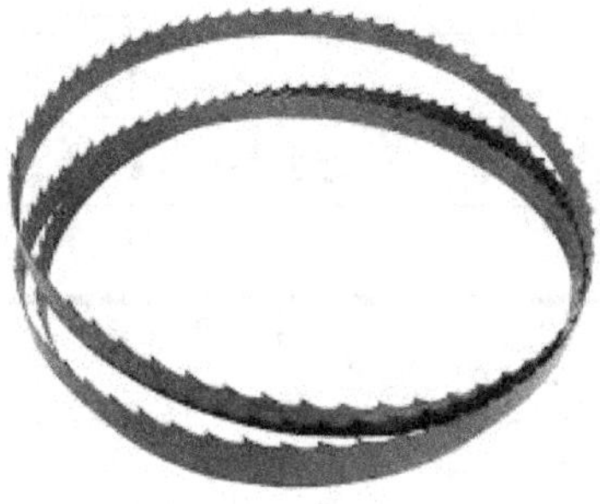   PT5 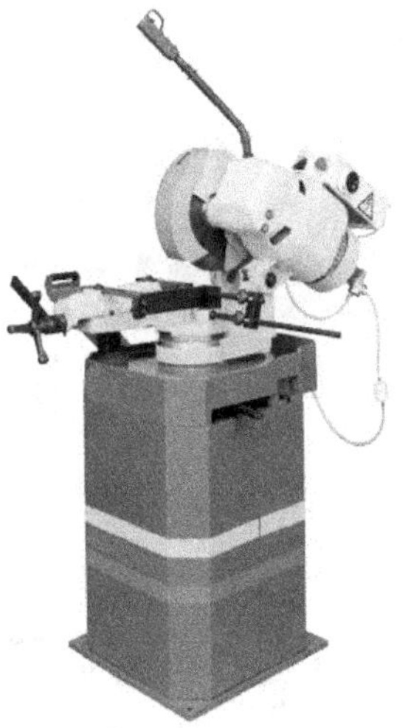   PT6

Part B:

Mark out, cut and drill the following shapes so they can be assembled.

12

38

29

3-Ø13 Holes

18

7- Equispaced Holes = 164

7-Ø3 Holes

18

82

200

Item 1 – 75 x 50 x 6 MS Bar

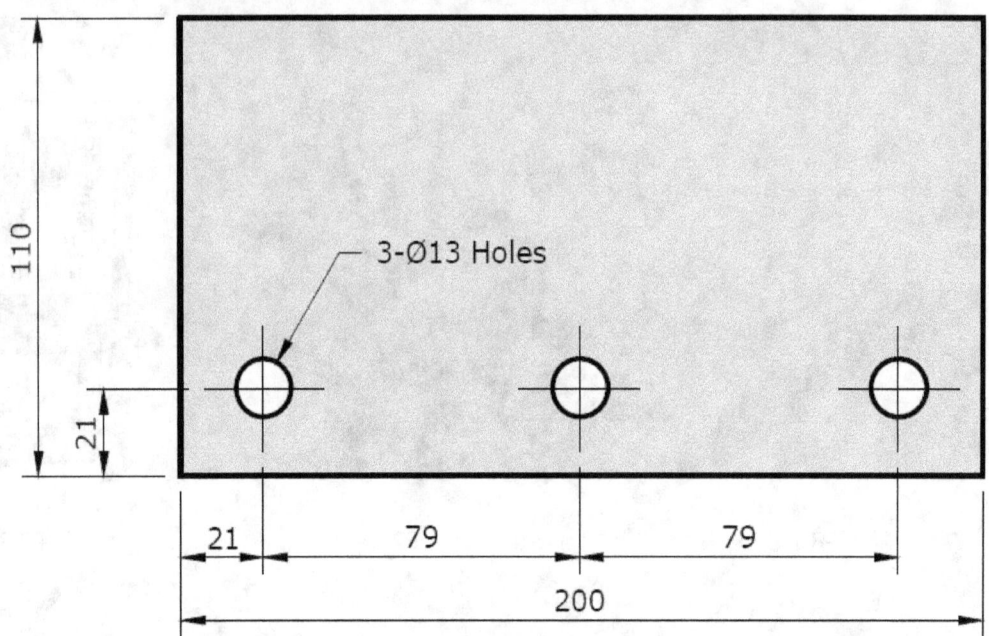

Item 2 – 15mm Mild Steel Plate

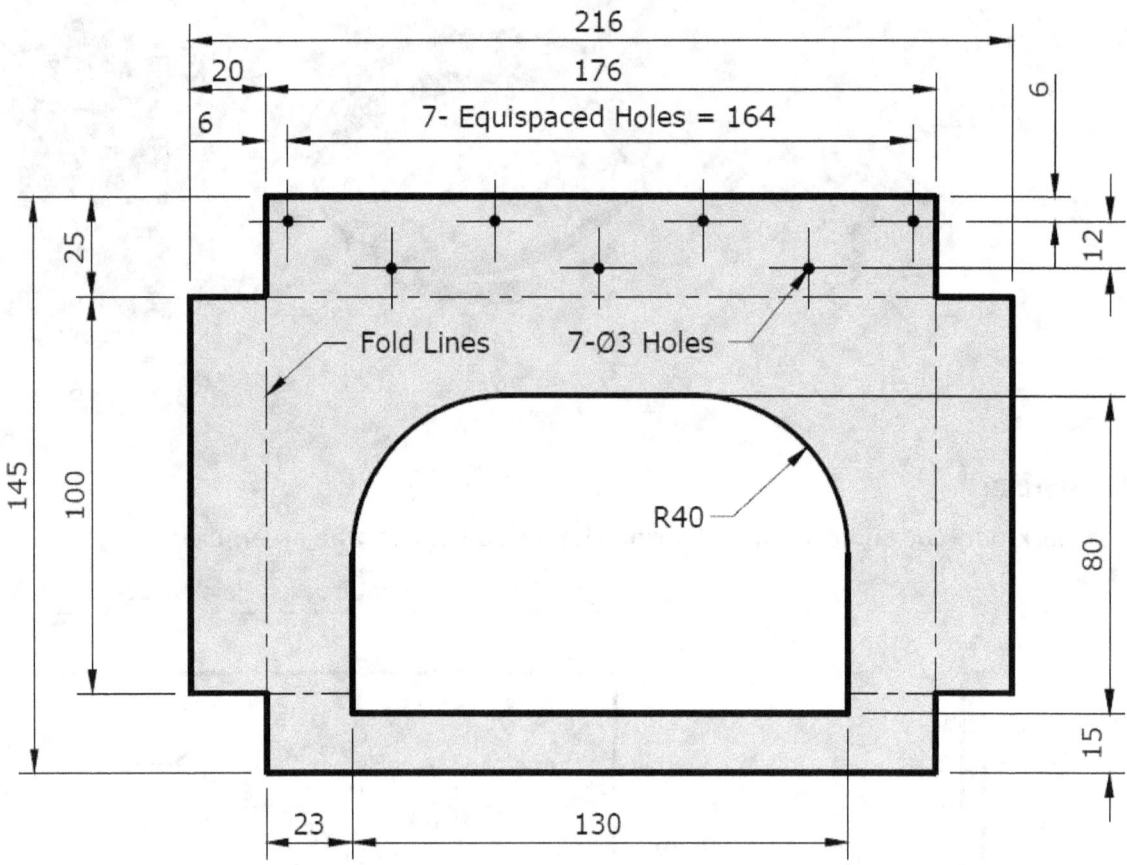

Item 3 – 1 mm Sheetmetal

Answers:

MEM05005-RQ-0201:
1. CNC Machine
2. D
3. Turret Lathe
4. Mill
5. Cost and skilled operators required.
6. C

MEM05005-RQ-0301:
1. A - Motor Tension Lever; B – Table; C – ON/OFF Power Switch; D – Raise/Lower Handles; E – Chuck; F – Locking Lever to Lower/Raise the Table; G – Spindle; H – Motor & Spindle Pulleys Inside the Cover; J – Pedestal Base.

2.	C	3.	True	4.	B	5.	A	6.	A
7.	D	8.	False	9.	C	10.	A	11.	B
12.	A	13.	C	14.	D				

15. A – Auger   B – Spade   C – High Speed D - Countersink

MEM05005-RQ-0401:
1. Power Shear
2. C
3. The front of the machine.
4. Bench Shears
5. B
6. A
7. Left Cut, Straight Cut & Right Cut.
8. D
9. Swing Shears
10. C
11. Throatless shears do not have handles.

MEM05005-RQ-0501:
1. Soluble oil base or Water-soluble synthetic.
2. C
3. Tungsten Carbide Tipped
4. B
5. D
6. C

MEM05005-RQ-0601:
1. A
2. A
3. D
4. B
5. C
6. D
7. B

MEM05005-PT-01:
1. A
2. Masonry Drill Bit.
3. 1/30th of the plate thickness.
4. High Speed Steel and Tungsten Carbide.
5. Alligator Shears.
6. Straight Cut.

Answers

7. 4,500 meters per minute.
8. A radial drill press is a machine tool that features an extended arm or beam along which a drill head can be moved to conveniently drill holes in large or cumbersome work pieces.
9. Gang Drill Press
10. 118°
11. A
12. C
13. A reamer is a precision cutting tool designed to finish a hole to a specific diameter.
14. Throatless Shear.
15. 18 TPI.
16. Cost and skilled operators required.
17. With the power turned OFF and when the saw blades have stopped turning.
18. Vibration is a major problem and even though the material is clamped in the vise there will always be varying thicknesses to cut.
19. teeth cannot be correctly matched
20. Counterbore bits form a flat bottom to the hole while countersink bits form 90° angles to the top surface of the hole.
21. Band Saw Blade.
22. Turret Lathe.
23. 10 mm.
24. Cold Saw.